Special Issue of the Manufacturing Engineering Society 2019 (SIMES-2019)

Special Issue of the Manufacturing Engineering Society 2019 (SIMES-2019)

Editors

Eva M. Rubio
Ana M. Camacho

MDPI • Basel • Beijing • Wuhan • Barcelona • Belgrade • Manchester • Tokyo • Cluj • Tianjin

Editors
Eva M. Rubio
Department of Manufacturing
Engineering, Universidad
Nacional de Educación a
Distancia (UNED)
Spain

Ana M. Camacho
Department of Manufacturing
Engineering, Universidad
Nacional de Educación a
Distancia (UNED)
Spain

Editorial Office
MDPI
St. Alban-Anlage 66
4052 Basel, Switzerland

This is a reprint of articles from the Special Issue published online in the open access journal *Applied Sciences* (ISSN 2076-3417) (available at: https://www.mdpi.com/journal/applsci/special_issues/society_2019).

For citation purposes, cite each article independently as indicated on the article page online and as indicated below:

LastName, A.A.; LastName, B.B.; LastName, C.C. Article Title. *Journal Name* **Year**, *Article Number*, Page Range.

ISBN 978-3-03943-621-7 (Hbk)
ISBN 978-3-03943-622-4 (PDF)

Contents

About the Editors

Eva María Rubio is a professor at the Department of Manufacturing Engineering of the National Distance Education University (UNED, Spain). She received her MSc in Aeronautical Engineering (1997) from the Polytechnic University of Madrid and her PhD in Industrial Engineering (2002) from the National University of Distance Education (UNED). She has developed her research activity in the field of manufacturing engineering, with the following main lines of work: analysis of machining processes; analysis of metal forming processes; industrial metrology; and teaching and innovation in engineering.

Ana María Camacho is a professor at the Department of Manufacturing Engineering of the National Distance Education University (UNED, Spain). She received her MSc in Industrial Engineering from the University of Castilla-La Mancha (UCLM) in 2001, and her PhD in Industrial Engineering from the UNED in 2005. Her main research interest is the innovation in manufacturing engineering and materials technology, especially focused on the analysis of metal forming and additive manufacturing techniques through computer aided engineering tools and experimental testing, and the development of methodologies for materials selection in demanding applications.

 *applied sciences*

Editorial

Special Issue of the Manufacturing Engineering Society 2019 (SIMES-2019)

Ana María Camacho * and Eva María Rubio

Department of Manufacturing Engineering, Industrial Engineering School, Universidad Nacional de Educación a Distancia (UNED), St/Juan del Rosal 12, E28040 Madrid, Spain; erubio@ind.uned.es
* Correspondence: amcamacho@ind.uned.es; Tel.: +34-913-988-660

Received: 24 February 2020; Accepted: 25 February 2020; Published: 27 February 2020

Abstract: The Special Issue of the Manufacturing Engineering Society 2019 (SIMES-2019) has been launched as a joint issue of the journals "Applied Sciences" and "Materials". The 10 contributions published in this Special Issue of Applied Sciences present cutting-edge advances in production planning, sustainability, metrology, cultural heritage, and materials processing with experimental and numerical results. It is worth mentioning how the topic "production planning" has attracted a great number of contributions in this journal, due to their applicative approach.

Keywords: additive manufacturing; 3D printing; forming; machining; metrology; production planning; technological and industrial heritage; industry 4.0; green manufacturing

---

After the complete success of the first edition [1] with 48 contributions on emerging methods and technologies, the Special Issue of the Manufacturing Engineering Society 2019 (SIMES-2019) [2] was launched as a joint issue of the journals "Applied Sciences" and "Materials".

Once again, this Special Issue was promoted by the Manufacturing Engineering Society (MES) [3] of Spain, with the aim of covering the wide range of research lines developed by the members and collaborators of the MES and other researchers within the field of Manufacturing Engineering.

In this Special Issue of the journal Applied Sciences, cutting-edge advances in production planning, sustainability, metrology, cultural heritage, and materials processing with experimental and numerical results have been published.

Concretely, the contributions have been mainly focused on the topics: additive manufacturing and 3D printing, with a contribution presenting the use of 3D printing with training purposes in the field of primary care [4]; advances and innovations in manufacturing processes, more specifically in the deep drawing of Inconel 718 applying different thermal treatments [5]; sustainable and green manufacturing, considering dry machining conditions in the turning of aluminum alloys used in aeronautical industry [6]; manufacturing of new materials such as a carbon fiber reinforced plastic (CFRP) laminates by drilling, using multiple sensor monitoring [7]; metrology and quality in manufacturing through the development of a bidimensional system for nanopositioning with uncertainty assessment [8]; manufacturing engineering and society, with a work presenting the use of hyperspectral imaging techniques with application in the conservation of cultural heritage [9]. Finally, it is worth mentioning how the topic "production planning" has attracted a great number of contributions in this journal due to their applicative approach, presenting the latest advances in methods with applications in the metallurgical [10], automotive [11], and military [12] industries, and involving innovative techniques such as machine learning and data mining [13].

After only three months since the publication of the first work [9], all the papers present prominent activity in their "article metrics", being remarkable how some of the papers belonging to this Special Issue have more than five hundred abstract and full-text views, which is clear evidence of the interest in all of these topics in readers of the journal Applied Sciences, in general, and scientists and professionals from the industry in particular.

Funding: This research received no external funding.

Conflicts of Interest: The authors declare no conflicts of interest.

References

1. Rubio, E.M.; Camacho, A.M. Special Issue of the Manufacturing Engineering Society (MES). *Materials* **2018**, *11*, 2149. [CrossRef] [PubMed]
2. Special Issue of the Manufacturing Engineering Society 2019 (SIMES-2019). Available online: https://www.mdpi.com/journal/materials/special_issues/SIMES_2019 (accessed on 20 February 2020).
3. Sociedad de Ingeniería de Fabricación. Available online: http://www.sif-mes.org/ (accessed on 20 February 2020).
4. Luque, M.C.; Calleja-Hortelano, A.; Romero, P.E. Use of 3D printing in model manufacturing for minor surgery training of general practitioners in primary care. *Appl. Sci.* **2019**, *9*, 5212. [CrossRef]
5. Ulibarri, U.; Galdos, L.; Sáenz de Argandoña, E.; Mendiguren, J. Experimental and Numerical Simulation Investigation on Deep Drawing Process of Inconel 718 with and without Intermediate Annealing Thermal Treatments. *Appl. Sci.* **2020**, *10*, 581. [CrossRef]
6. Martín-Béjar, S.; Javier, F.; Vilches, T.; Gamboa, C.B.; Hurtado, L.S. Cutting Speed and Feed Influence on Surface Microhardness of Dry-Turned UNS A97075-T6 Alloy. *Appl. Sci.* **2020**, *10*, 1049. [CrossRef]
7. Teti, R.; Segreto, T.; Caggiano, A.; Nele, L. Smart Multi-Sensor Monitoring in Drilling of CFRP/CFRP Composite Material Stacks for Aerospace Assembly Applications. *Appl. Sci.* **2020**, *10*, 758. [CrossRef]
8. Díaz-Pérez, L.; Torralba, M.; Albajez, J.A.; Yagüe-Fabra, J.A. 2D positioning control system for the planar motion of a nanopositioning platform. *Appl. Sci.* **2019**, *9*, 4860. [CrossRef]
9. Bayarri, V.; Sebastián, M.A.; Ripoll, S. Hyperspectral imaging techniques for the study, conservation and management of rock art. *Appl. Sci.* **2019**, *9*, 5011. [CrossRef]
10. Gejo García, J.; Gallego-García, S.; García-García, M. Development of a Pull Production Control Method for ETO Companies and Simulation for the Metallurgical Industry. *Appl. Sci.* **2020**, *10*, 274. [CrossRef]
11. Gallego-García, S.; Reschke, J.; García-García, M. Design and simulation of a capacity management model using a digital twin approach based on the viable system model: Case study of an automotive plant. *Appl. Sci.* **2019**, *9*, 5567. [CrossRef]
12. Acero, R.; Torralba, M.; Pérez-Moya, R.; Pozo, J.A. Value stream analysis in military logistics: The improvement in order processing procedure. *Appl. Sci.* **2020**, *10*, 106. [CrossRef]
13. Qiu, Y.; Ji, W.; Zhang, C. A hybrid machine learning and population knowledge mining method to minimize makespan and total tardiness of multi-variety products. *Appl. Sci.* **2019**, *9*, 5286. [CrossRef]

applied sciences

Article

2D Positioning Control System for the Planar Motion of a Nanopositioning Platform

Lucía Díaz-Pérez [1,*], **Marta Torralba** [2], **José Antonio Albajez** [1] **and José Antonio Yagüe-Fabra** [1]

1   I3A—Universidad de Zaragoza, C/María de Luna, 3, 50018 Zaragoza, Spain; jalbajez@unizar.es (J.A.A.); jyague@unizar.es (J.A.Y.-F.)
2   Centro Universitario de la Defensa Zaragoza, Ctra. Huesca s/n, 50090 Zaragoza, Spain; martatg@unizar.es
*   Correspondence: lcdiaz@unizar.es

Received: 24 October 2019; Accepted: 8 November 2019; Published: 13 November 2019

Abstract: A novel nanopositioning platform (referred as NanoPla) in development has been designed to achieve nanometre resolution in a large working range of 50 mm × 50 mm. Two-dimensional (2D) movement is performed by four custom-made Halbach linear motors, and a 2D laser system provides positioning feedback, while the moving part of the platform is levitating and unguided. For control hardware, this work proposes the use of a commercial generic solution, in contrast to other systems where the control hardware and software are specifically designed for that purpose. In a previous paper based on this research, the control system of one linear motor implemented in selected commercial hardware was presented. In this study, the developed control system is extended to the four motors of the nanopositioning platform to generate 2D planar movement in the whole working range of the nanopositioning platform. In addition, the positioning uncertainty of the control system is assessed. The obtained results satisfy the working requirements of the NanoPla, achieving a positioning uncertainty of ±0.5 µm along the whole working range.

Keywords: 2D positioning control; nanopositioning; Halbach linear motors; positioning uncertainty

---

1. Introduction

In recent years, the applications of nanotechnology and nanoscience have increased, demanding high accuracy positioning systems capable of working in large ranges at the nanometre scale. These positioning stages can be used for measuring or nanomanufacturing applications by integrating different devices [1]. The performance of these processes is directly related to the accuracy of the positioning systems and their working range [2]. Therefore, accurate positioning control in a large working range is one of the main necessities of nanotechnology applications [3].

At the University of Zaragoza, a novel 2D nanopositioning platform (NanoPla) is in development [4]. This NanoPla has been designed to work together with different kinds of tools and probes in various applications, such as metrology or nanomanufacturing. In particular, the main application of this first prototype is surface topography characterisation at the atomic scale of samples with relatively large planar areas. The measuring device will be attached to a moving platform that performs a large displacement of 50 mm × 50 mm. The NanoPla architecture is based on a scheme of the four integrated custom-made Halbach linear motors, which allow the implementation of planar motion [5]. The XY position of the platform is measured by a 2D plane mirror laser interferometer system. The design of the NanoPla has been optimised to achieve a nanometre resolution along its whole working range [6].

A commercial control solution for these custom-made linear motors is not currently available. Thus, in other previous works [7,8], control hardware and software were specifically designed and built for control issues. In contrast, in this project, the use of commercial control hardware for generic motors has been proposed as a novel solution for their control and drive systems. The purpose of this research is to facilitate the future replicability of the system. In a previous study [9], a 1D positioning control strategy was designed and implemented in the selected commercial control hardware for one linear motor performing movement on a linear guide. The right performance of this control system, according to the established design requirements, was experimentally verified. In this work, the control strategy is optimised for 2D movement and implemented in the NanoPla to control and drive the planar motion of a nanopositioning platform in a range of 50 mm × 50 mm. Subsequently, the positioning uncertainty of the proposed control system is analysed.

This article is an extension of the work presented in a previous conference paper [10] and is divided as follows. Firstly, an overview of the NanoPla is presented, and the Halbach linear motors, the positioning sensor, and the hardware of the control system are described. Secondly, a positioning 2D control system is proposed and experimentally validated. Then, the positioning uncertainty of the 2D positioning control system is assessed. Finally, conclusions are developed.

2. Materials and Methods

This section describes the NanoPla's design and its applications. Additionally, the control system's hardware is defined, as well as the actuators, the positioning sensor, and the connections between them.

2.1. 2D Nano-Positioning Platform (NanoPla)

The NanoPla design was presented in [6]. An exploded view of the NanoPla can be seen in Figure 1. This platform has a three-layered structure that consists of fixed inferior and superior bases and a moving platform that is placed between them. The metrology loop consists of two metrology frames: One in the moving platform (I) and the other in the inferior base (II). The structural parts of the NanoPla are made of the aluminium alloy 7075-T6 due to its strength, which is comparable to many steels. In this first prototype for preliminary tests, the metrology frame is made of the same aluminium. Nevertheless, the selected material for its final design is Zerodur, due to its low thermal expansion coefficient that assures negligible dimensional changes caused by temperature variation.

The moving platform is levitated by three vacuum-preloaded air bearings, while four Halbach linear motors perform its motion. A Halbach linear motor has two parts: A permanent magnet array and a stator that consists of three-phase ironless coils. In the NanoPla, the magnet arrays of the four linear motors are fixed to the moving platform, and the stators are assembled to the superior base, which minimises the weight of the moving part. A 2D laser interferometer system works as a positioning sensor. The laser heads are positioned in the inferior base, and the positioning mirrors are fixed to the moving platform. Out-of-plane deviations are characterised by three capacitive sensors. As a probe, the aim is to embed an atomic force microscope (AFM) in the NanoPla. The AFM is fixed to the moving platform that positions it in the XY-plane above the certain area of the sample to be measured, allowing the characterization of a large area of the sample (50 mm × 50 mm). Once the AFM is positioned, the moving platform remains static (air bearings off) in order to carry out the scanning task.

The NanoPla offers a two-stage scheme, that is, the XY-long range positioning of the moving platform (coarse motion) is complemented by an additional piezo-nanopositioning stage that is fixed to the metrology frame of the inferior base (fine motion). This second stage is a commercial piezo-nanopositioning device specifically designed for the scanning probe and optical microscopy (model NPXY100Z10A from nPoint). It has a XYZ working range of 100 μm × 100 μm × 10 μm and a positioning noise of 0.5 nm in the XY-plane and 0.1 nm in Z-axis. During the scanning operation, the piezostage will perform the motion of the sample. Therefore, the position control system should have a positioning error at least one order of magnitude smaller than the maximum XY range of the commercial piezo-nanopositioning stage (i.e., 10 μm). Thus, this error could be corrected by the fine motion of the piezo stage.

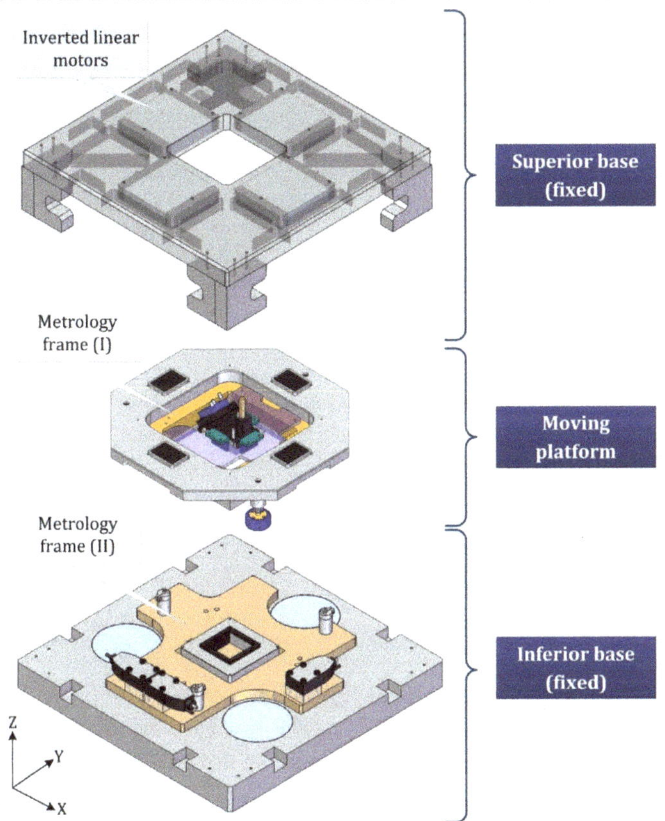

Figure 1. Exploded view of the nanopositioning platform (NanoPla).

2.2. Components of the Control System

The positioning systems' main components are actuators, control hardware, and positioning sensors. The selection of these components is based on the positioning stage's required precision, as well as the operating range and structure of the stage. Similarly, the development of the control strategy is constrained by the positioning system components defined by the NanoPla's design. In addition, the implementation of the components in the positioning control system can be optimised to leverage the capabilities of each of component in order to obtain the maximum positioning accuracy.

Halbach linear motors directly transform electrical energy into linear motion and have been selected as actuators in the NanoPla because of their many advantages in precision engineering due to their lack of mechanical transmission elements (thereby avoiding, for example, backlash). Similarly, contactless unguided motion prevents friction and allows planar motion. Planar motion is preferred in precision applications because it minimises geometrical errors and presents many other advantages in precision engineering [11]. The Halbach linear motors used in the NanoPla were developed by Trumper et al. [5] and are not commercialised. Therefore, they have been custom-made at the University of North Carolina at Charlotte, and the size of their winding areas is large enough to allow planar movement in the 2D working range of the NanoPla. When DC current flows through the three-phase coils of a Halbach linear motor, the electric field interacts with the magnetic field of the magnet array, resulting in two orthogonal forces—one horizontal and the other vertical. The relationship between the phase currents and the generated forces is defined by the motor law presented in [9]. The vertical forces of the four motors facilitate the moving platform's levitation, while the horizontal forces move the platform in the XY-plane.

A generic commercial control system for custom-made Halbach linear motors is not currently available. For this reason, other reviewed stages integrating these motors specifically designed and developed control hardware and software for this purpose [7,8]. In these stages, control strategies are based on individually and independently controlling each of the phase currents of the motors with transconductance power amplifiers. However, following the NanoPla's design principle of integrating commercial devices when possible, the use of only commercial hardware and no custom-made electronics was proposed. Thus, a Digital Motor Control Kit (DMC) DRV8302-HC-C2-KIT from Texas Instruments was selected as the control hardware. The DMC kit consists of a F28035 control card and a DRV8302 board. This DMC kit has been designed for rotary brushless DC and permanent magnet synchronous three-phase motors, where the aim is to control the rotation speed or the torque generated. The board includes a three-phase power stage that drives and generates phase voltages by pulse-width modulation (PWM), in contrast to other works [7,8] where the hardware acted as a controlled current source. Additionally, the control hardware forces the star connection of the phases, thereby impeding the phase currents from being controlled independently, as was done in [7,8]. This hardware has been selected due to its relatively low cost and the advantages of the associated software. The use of the microcontroller from Texas Instruments is based on the Target Support Package™ for Embedded Code. This Package integrates MATLAB® and Simulink® with Texas Instrument tools and C2000 processors to generate, compile, implement, and execute the optimised control code with a user-friendly graphic interface and without programming in a specific language.

As mentioned, in the NanoPla, a 2D laser system is used as a positioning sensor to provide control system feedback. This 2D laser system is a combination of three 1D plane mirror laser interferometer systems. Laser interferometer systems provide highly accurate, non-contact measurements and are capable of working at long distances [12]. In addition, the use of plane mirrors as retroreflectors allows one to measure planar motion [13]. Two laser beams are needed to measure the displacement in the X- and Y-axes. In addition, one more beam is needed to determine the rotation around Z-axis. Thus, one laser head is placed projecting its beam in the X-axis, aligned with the reference system of the travel range, while the other two laser heads project their beams parallel to the Y-axis. The laser system components belong to the Renishaw RLE10 laser interferometer family, which consists of a laser unit (RLU), three sensor heads (RLD), two plane mirrors (one per axis), and an environmental control unit (RCU). Furthermore, an external interpolator is used to reduce the expected resolution of the system from 9.88 nm to 1.58 nm. Besides the readouts of the three laser encoders, the system also provides the readouts of the RCU sensors: Air temperature, material temperature, and air pressure. The measurement of each signal takes approximately 0.04 s; thus, the maximum speed at which it is possible to record the six measurements is every 0.25 s.

In a previous work [9], an experimental setup, external to the NanoPla, was assembled for the development and experimental validation of the control system of one Halbach linear motor in 1D. In that setup, the magnet array of the motor was fixed, and the stator of the motor was the moving part, attached to a pneumatic linear guide. A control strategy was implemented in the DMC kit, and a 1D interferometer laser system was used as the positioning sensor. After that first validation, the four linear motors were installed in the NanoPla, and each of them connected to one DMC Kit for the project presented in this paper. In turn, all the DMC kits were connected to the host PC that coordinated the four motors. Movement was achieved while the moving platform was levitated by three air bearings. Figure 2a is a photograph of the NanoPla and the control system components, while Figure 2b illustrates a scheme of the connections between the host PC, the positioning sensors, the control hardware, and the linear motors. The input is the target position in the X, Y coordinates, which is entered by the user at the host PC. The control strategy is computed by the PC that receives the position feedback from the laser system. Then, the PC computes the phase voltages that must be generated for the control hardware to drive the linear motors that produce the movement. The plane mirrors are the moving target of the 2D laser system, and the magnet arrays are the part of the linear motors that perform the relative movement respective to the stator that is fixed. The plane mirrors, as well as the magnet arrays, belong to the moving platform.

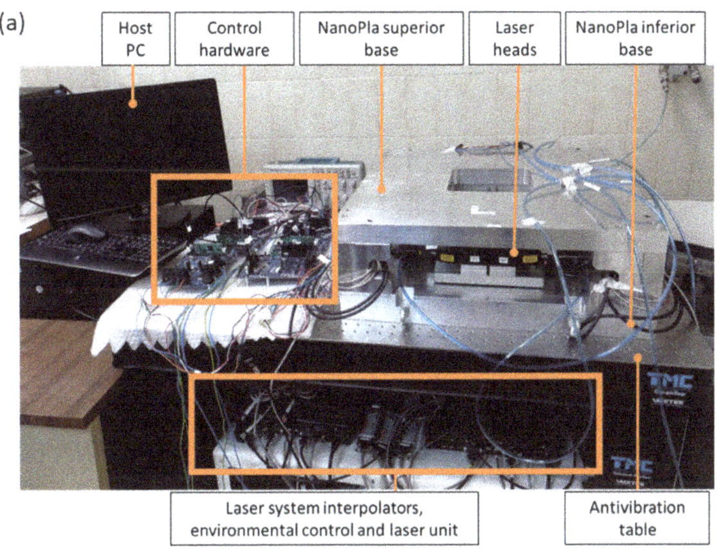

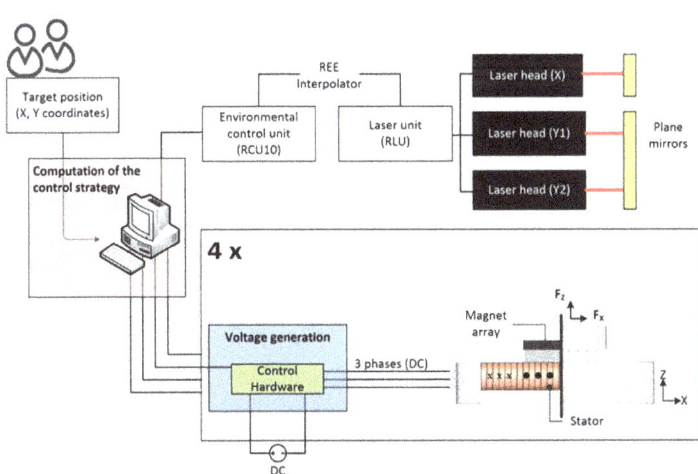

Figure 2. (**a**) Photograph of the NanoPla and the control system components and setup: Host PC, control hardware, and 2D laser system units; (**b**) scheme of the connections between the host PC, control hardware, and positioning sensor in the 2D positioning control system.

3. 2D Positioning Control

After presenting the main components and scheme of the 2D control system, this section first analyses the NanoPla dynamic model and then presents a 2D positioning control strategy for the NanoPla. Finally, the NanoPla's performance is experimentally validated.

3.1. Dynamic Characterisation of the System

In the NanoPla, the motors are placed in parallel pairs (Figure 3); thus, two motors, motor 1 and motor 2 (represented as M1 and M2 in Figure 3b), generate forces on the X-axis (F_{M1} and F_{M2}) that move the platform in the X-direction. Similarly, the other two parallel motors, motor 3 and 4 (M3 and M4), generate forces on the Y-axis (F_{M3} and F_{M4}) that move the platform in the Y-direction. In addition, the four motors are placed symmetrically at a distance R (169.9 mm) from the centre of the platform and, thus, their forces generate a torque at the centre of the moving platform, around the Z-axis. The movements of the platform in the X- and Y-axes -X_s and Y_s- and the rotation around Z-axis -θ_{zs}- are monitored by the 2D laser interferometer system (Laser Y1, Y2, and X). The total forces in the X- and Y-axes and the torque around the Z-axis -T_z- can be calculated as follows:

$$F_x = F_{M1} + F_{M2} \tag{1}$$

$$F_y = F_{M3} + F_{M4} \tag{2}$$

$$T_z = -F_{M1}\cdot R + F_{M2}\cdot R - F_{M3}\cdot R + F_{M4}\cdot R \tag{3}$$

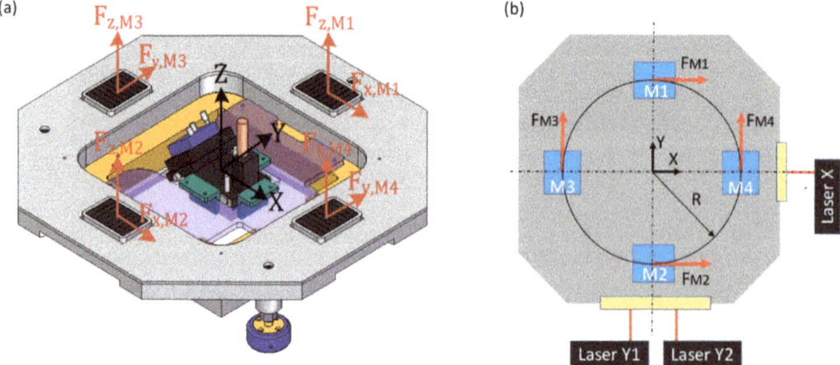

Figure 3. Scheme of the forces that act on the moving platform: (**a**) isometric view; (**b**) top view scheme.

The dynamic model of a Halbach linear motor working as a positioning actuator in a pneumatically levitated linear stage was identified as a servosystem in a previous work [14]. In this system, the electromagnetic horizontal force generated by the motor around the stable equilibrium position ($F_x = 0$, $F_z > 0$) acts as a proportional controller. The closed-loop transfer function that relates the equilibrium position, $X_{eq}(s)$, to the position of the stage, $X(s)$, was identified with a spring-mass-damper model (second order system). The NanoPla 2D positioning model is expected to present a similar second-order transfer function in each axis of motion, since it uses the same actuators and the moving platform is also pneumatically levitated. The transfer function of the system can be obtained experimentally, which enables a better understanding of the system and allows tuning of the controllers in advance, which facilitates tasks in the experimental setup.

The force generated by each Halbach motor around the stable equilibrium position (linear zone) can be defined as

$$F_{M,x}(s) = K_M(X_{eq}(s) - X(s)) \tag{4}$$

where K_M is the slope of the thrust force generated by each Halbach linear motor around the equilibrium position. The forces along the X-axis generated by the parallel motor pair M1 and M2 at the initial position are represented in Figure 4. The linear zone and the slope around the stable equilibrium position are also represented in Figure 4. As shown, the linear zone has a length of approximately 5 mm, with the stable equilibrium position in the middle of the region. In this system, the stable equilibrium position of the parallel pair of Halbach linear motors M1 and M2 is set at the same X-coordinate. Thus, the total force, F_X, generated around the stable equilibrium position is twice the force defined in Equation (4). Therefore, the transfer function that relates the final position of the stage X(s) with the defined stable equilibrium position X_{eq}(s) is the following:

$$\frac{X(s)}{X_{eq}(s)} = \frac{\frac{2K_M}{m}}{s^2 + \frac{b_X}{m}s + \frac{2K_M}{m}} \tag{5}$$

where m is the mass of the moving part and b_X defines the viscous-friction elements of the setup and the eddy-current damping of the motors in the X-axis. Additionally, $2K_M$ is the slope of the total thrust force, F_X, generated by the pair of Halbach linear motors around the equilibrium position in the X-axis (Figure 4). Since the actuator distribution in the moving platform is symmetrical, the same equations are considered valid for the Y-axis. The mass of the moving platform is known (13.25 kg), and the value of K_M depends on the reference value set for the vertical force at the stable equilibrium position, F_{zref}. This value defines the amplitude of the sinusoidal distribution of the forces along the axis of motion, while the spatial period is defined by the design [9]. For instance, when F_{zref} is set to 1 N, K_M is approximately 205 N/m, and when F_{zref} is set to 2 N, K_M is 410 N/m. Therefore, K_M and m are known, whereas the viscous-friction factor b_X must be obtained experimentally.

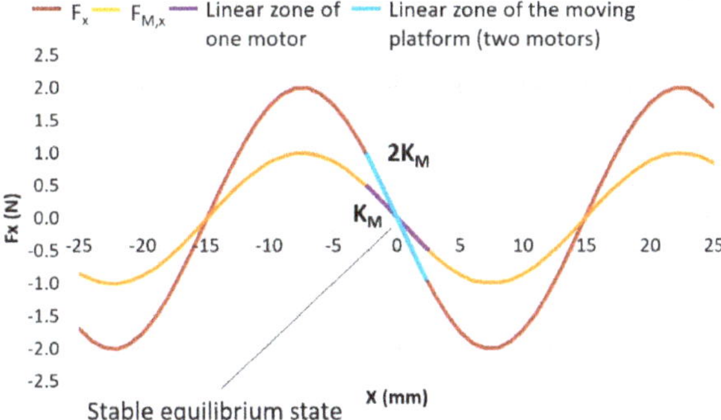

Figure 4. Generated forces by one ($F_{M,x}$) and two motors (F_x) along the X-axis of movement.

On the other hand, rotation around the Z-axis (θ_z) is generated by torque acting on the central point of the moving platform. This torque (T_z) is the sum of the torques generated by each motor in the plane of motion (Figure 3b). The 2D plane mirror laser interferometer system requires the rotation of the moving platform to be less than $\pm 1.2 \times 10^{-4}$ rad, according to the manufacturer's specifications, to prevent misalignment between the laser beam and plane mirror. Therefore, to rotate the moving platform by angle $\Delta\theta_z$, each motor would have to perform a displacement of the same magnitude, equal to $\Delta\theta_z \cdot R$, in the direction that favours that rotation. According to this, the torque generated by the four motors of the NanoPla to perform a rotation of $\Delta\theta_z$ can be calculated as follows:

$$T_z = 4K_M R^2 \Delta\theta_z \tag{6}$$

It is worth mentioning that the angular position θ_z is considered to be a stable equilibrium angular position when, at that angular position, the platform remains still and after a disturbance, it comes back to the same angular position. This stable equilibrium position is created when the moving platform is perfectly aligned in the X- and Y-axes—that is, $\theta_{z,eq} = 0$. Thus, as in the previous case, the transfer function that relates the final angular position of the stage θ_z(s) with the defined stable equilibrium angular position $\theta_{z,eq}$(s) is defined by Equation (7), where b_θ is the damping factor in θ_z, and I_z is the inertia of the moving platform.

$$\frac{\theta_z(s)}{\theta_{z,eq}(s)} = \frac{\frac{4K_M R^2}{I_z}}{s^2 + \frac{b_\theta}{I_z}s + \frac{4K_M R^2}{I_z}}$$ (7)

3.2. 2D Control Strategy

The four motors are placed in parallel pairs. Thus, as long as the motor pairs displace the same distance, the moving platform remains aligned. It was experimentally verified that the moving platform can be positioned along its working range by controlling each of the motors individually with a 1D positioning strategy (presented in [9]) implemented in each motor. Nevertheless, independently controlling each motor results in undesired rotations around the Z-axis during the transient response—that is, when the platform is moving from one position to other. Undesired rotations cause misalignments between the laser beams and plane mirrors, which can affect the performance of the laser system and even impede the displacement measurements. Therefore, in order to prevent undesired rotations of the moving platform during motion, it is necessary to coordinate the control of the four motors in a 2D control strategy.

On the other hand, in [7], control of the out-of-plane motion was proven to be unnecessary due to the high stiffness of the air bearings. Similarly, in the NanoPla, the moving platform is levitated by three vacuum-preloaded air bearings with an input pressure of 0.41 MPa and an input vacuum of 15 mmHg, as recommended by the manufacturer. At these working conditions, the air bearings have a stiffness of 13 N/μm. As calculated in [9], the variation of the vertical force during motion has a maximum increment of 7%, which results in negligible vibrations of the moving platform. Therefore, control of the out-of-plane motion in the NanoPla is unnecessary, although it will be monitored by three capacitive sensors.

The proposed control strategy positions the platform in the X- and Y-axes and minimises rotations around the Z-axis ($\theta_{zref} = 0$), to prevent laser system misalignments. This is done with three independent proportional-integral-derivative (PID) controllers that act on the forces generated in the X- and Y-axes by the motor pairs (F_x and F_y) and on the torque (T_z) generated by the four motors. The positioning feedback (X_s, Y_s, and θ_s) is provided by the three laser beams (Laser Y1, Y2, and X) of the laser system. Considering the symmetry of the moving platform, the total forces and torque are divided between the four motors, and the horizontal force that each of the motors needs to generate is computed.

Figure 5 illustrates the scheme of the control system that has been implemented in this project. The input of the control system is the target position (X_{ref}, Y_{ref}) of the moving platform that is entered in the graphic user interface. In addition, the rotation around the Z-axis should be kept minimal ($\theta_{zref} = 0$). The previously described control strategy is computed in Simulink® on the host PC (Figure 5). Firstly, this strategy calculates the horizontal forces that each of the linear motors needs to generate to correct positioning errors. Then, the phase currents that each linear motor requires to generate those forces are calculated according to the commutation law defined in [9]. The outputs of the control strategy are the corresponding phase voltages that the control hardware must generate for each motor. These phase voltages are generated at the power stage of each DMC kit. Then, the interaction of the phase currents flowing through the stator coils with the magnetic field of the magnet arrays of the moving platform generates horizontal forces that move the platform. This movement is recorded by the laser system and fed back to the control strategy.

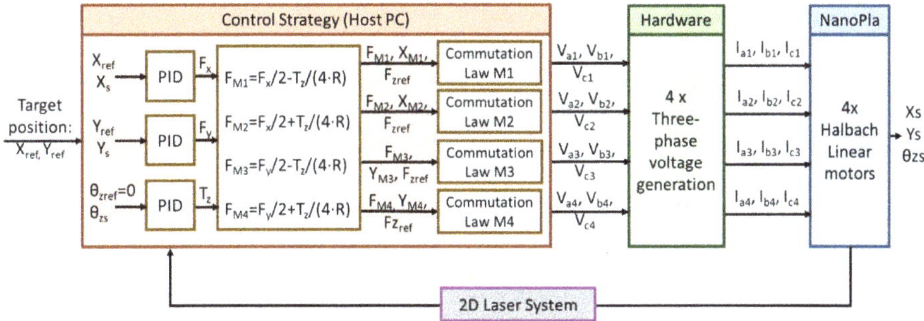

Figure 5. Scheme of the 2D position control system.

3.3. Experimental Results

The procedure for experimentally obtaining the transfer function in a linear stage was defined in [14]. This procedure has been adapted to the NanoPla, where a parallel pair (M1 and M2) generates the thrust force in the X-axis, whereas the other parallel pair (M3 and M4) generates the force in the Y-axis (Figure 3). In order to obtain the transfer function in the X-axis, the motor pairs aligned in the Y-axis are set to remain still at the initial position in the Y-axis ($Y_{ref} = 0$, $\theta_{z,ref} = 0$), acting as a guiding system that prevents the movement of the moving platform on the Y-axis and its rotation around the Z-axis. Then, the NanoPla is displaced from its initial position inside the linear zone by using the electromagnetic force of the motor pair aligned on the X-axis (M1 and M2). The movement of the moving platform is recorded by the laser system, and then the spring mass damper model of Equation (5) is fit to the response. The same procedure is followed to obtain the transfer function on the Y-axis.

Figure 6 represents the experimentally obtained 1 mm step response of the stage on the X- and Y-axes (X_s and Y_s) when F_{zref} is set to 2 N. The simulated response of the adjusted model is also shown (X_{sim} and Y_{sim}). The obtained values for m and K_M match the actual mass of the moving platform and the slope of the thrust force around the stable equilibrium position, respectively. The viscous-friction elements (b) have a value of 56.8 N·s/m on the X-axis and for 63.07 N·s/m on the Y-axis. The differences between axes could be due to geometrical errors in assembly and the fact that the moving platform is not perfectly symmetrical due to the presence of the plane mirrors, which are larger for the dual beam Y-axis. In addition, it was experimentally verified that by varying F_{zref}, the value of K_M changed as expected, and the values of m and b remained invariable.

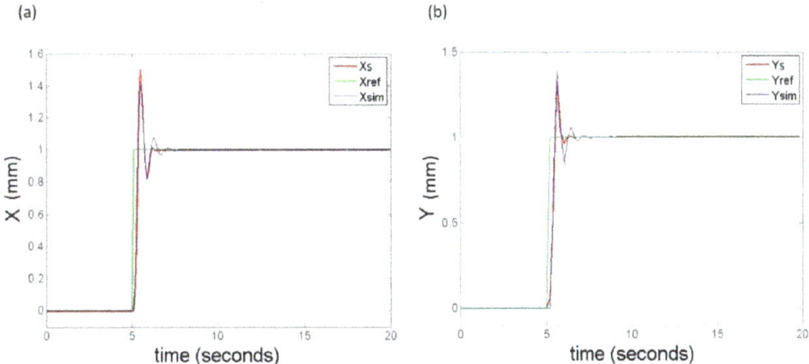

Figure 6. (a) The 1 mm step response on the X-axis of the stage in the open loop and simulation of the plant; (b) the 1 mm step response on the Y-axis of the stage in the open loop and of the simulation of the plant.

In order to obtain the transfer function for the rotation around the Z-axis, the four motors are set to displace from an initial rotated position to a stable equilibrium position $\theta_{z,eq} = 0$. At this initial position, the angular deviation is not null. In order to define the initial position, the four motors are displaced at a distance of $\Delta\theta_z \cdot R$, equal to 15 µm. In each motor pair, each motor displaces in an opposite direction to contribute to the rotation around the central point of the stage, generating an angular deviation of 8.83×10^{-5} rad, which is close to the maximum displacement allowed without losing the laser system's alignment. The experimentally obtained values for I_z and K_M approximately match the actual inertia of the moving platform and the slope of the thrust force around the stable equilibrium position, respectively. Nevertheless, due to the limited range of the angular deviation and the short response time, the recorded response is not smooth enough to perfectly match the simulated plant. Figure 7 represents the experimentally obtained 8.83×10^{-5} rad step response of the stage around the Z-axis (θ_s). The simulated response of the adjusted model is also shown (θ_{sim}).

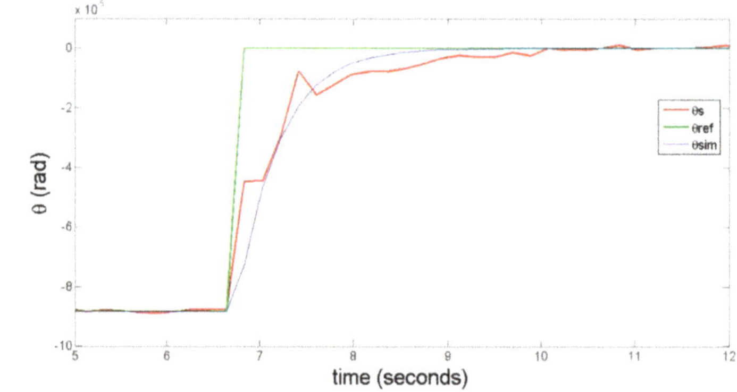

Figure 7. Angular step response of 8.83×10^{-5} rad around the Z-axis of the stage in the open loop and simulation of the plant.

After the transfer function identification, the proposed control system was implemented in the NanoPla, and its correct performance was experimentally verified. For these experiments, the vertical force generated by each motor was defined as 2 N, which limits the phase currents' working range to ±0.83 A.

Firstly, the stability of the positioning control system was examined for a long period of time. The moving platform was set to remain still at the initial position for 30 min, and it was experimentally verified that the position deviations are confined to ±1 µm. In Figure 8, a 400 s period of this test is shown, and the root mean square (RMS) deviation during this time is 0.28 µm. The results in the Y-axis are similar, since the moving platform is symmetrical. During this period of time, the force generated by each linear motor varied between ±2 mN—that is, the system worked around the stable equilibrium position, as expected.

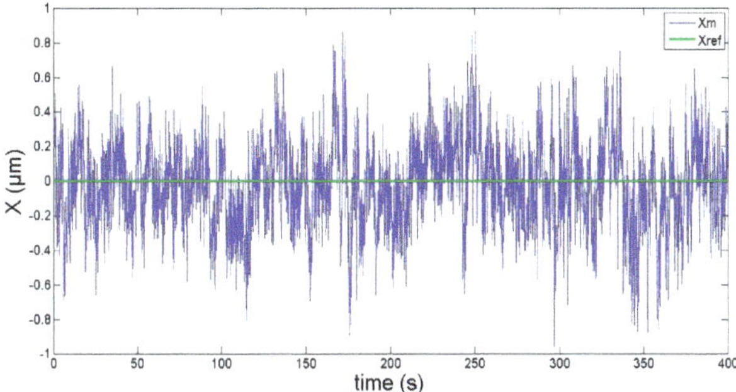

Figure 8. Long-term stability analysis of the developed positioning control system.

As mentioned earlier, the NanoPla has a two-stage scheme. The moving platform performs coarse movement in a large working range of 50 mm × 50 mm. Once the moving platform arrives at the target position, it stays static (air bearings off), and a piezo-nanopositioning stage placed on the inferior base performs the fine displacement required for the scanning task. The selected commercial piezostage has a working range of 100 μm × 100 μm in the XY-plane. Therefore, as mentioned, it is defined as a requirement that the position control system has a positioning error smaller than 10 μm, so this error can be corrected by the fine motion of the piezo stage.

Therefore, the performance of the control system was tested when performing a displacement to the target position. Then, 10 μm step responses were taken in the X- and Y-directions. At the same time, the perturbation to the other axis was also recorded, as shown in Figure 9. The perturbed motions in the other axis demonstrate that there is a dynamic coupling between axes. This is unavoidable because there is only one moving part that is affected by the vibrations of the four motors. This perturbation generates a displacement on the Y-axis of a maximum of 2 μm during the transient response. Nevertheless, once the NanoPla achieves its target position on the X-axis, the positioning error on the Y-axis is corrected. In addition, it has been observed that during the transient response, the maximum angular deviation is 1.5×10^{-5} rad, which is inside the tolerance of $\pm 1.2 \times 10^{-4}$ rad required for the laser system to read.

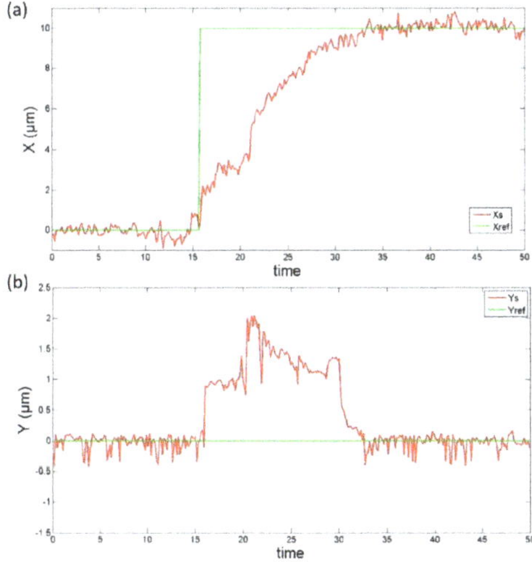

Figure 9. The 10 μm step response on the X-axis (**a**) and perturbation on the Y-axis (**b**).

Similarly, 100 μm step responses were taken in the X- and Y-directions, while the perturbation to the other axes was also recorded, as shown in Figure 10.

As in the previous case, the displacement in one axis generates perturbations in the other axis. This perturbation generates a displacement of a maximum of 7 μm on the Y-axis during the transient response, which is corrected when the NanoPla achieves its stationary state. In addition, it has been observed that during the transient response, the maximum angular deviation is 7.3×10^{-5} rad, which is inside the tolerance of $\pm 1.2 \times 10^{-4}$ rad required for the laser system to read.

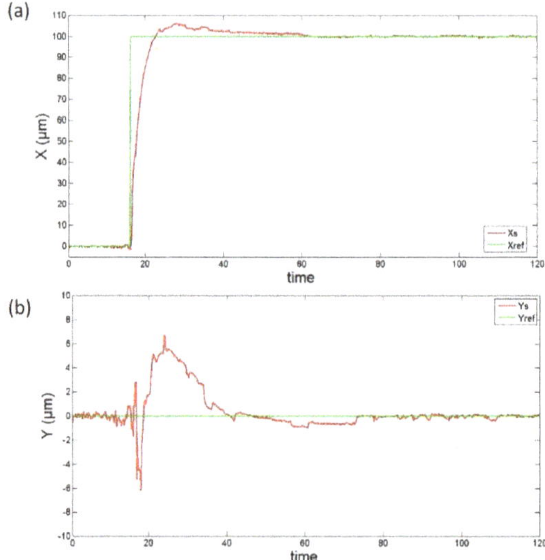

Figure 10. The 100 μm step response on the X-axis (**a**) and the perturbation on the Y-axis (**b**).

It should be taken into account that the transient response can be adjusted depending on the requirements of the application by changing the parameters of the PID controller.

On the other hand, planar scanning motion is the typical motion used in precision engineering such as nanomanufacturing and metrological characterisation. Several experimental results are presented to demonstrate the scanning capability of the developed positioning control system. Figure 11a shows a displacement on the X-axis of the moving platform from the centre to one extreme of the working range at a constant speed. Similarly, Figure 11b shows a 1 mm forward and backward displacement on the Y-axis.

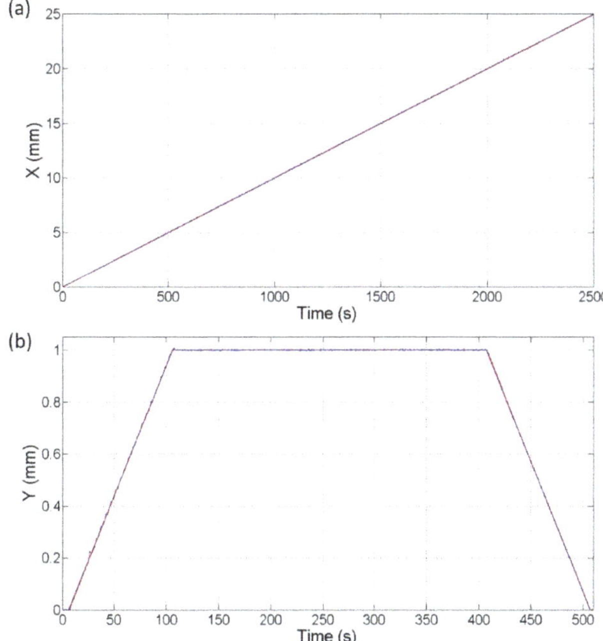

Figure 11. (**a**) The 25 mm displacement at a constant speed on the X-axis; (**b**) the 1 mm forward and backward displacement at a constant speed on the Y-axis.

In addition, it has been verified that the platform can perform simultaneous movement on the X- and Y-axes without losing the alignment between the laser beam and plane mirrors. Figure 12 illustrates the circular motion performed simultaneously on the X- and Y-axes.

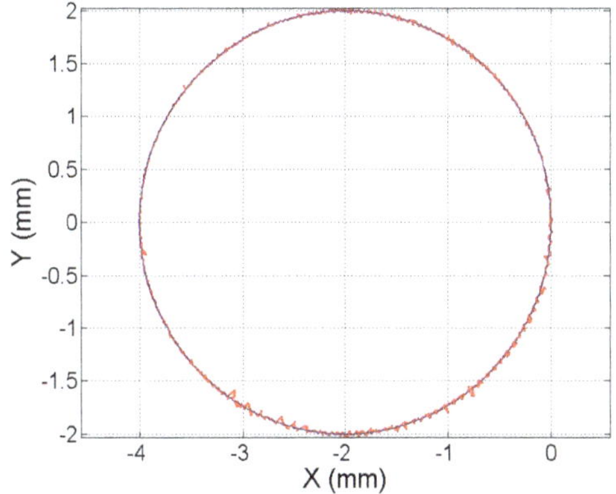

Figure 12. Circular motion performed simultaneously on the X- and Y-axes.

4. Positioning Uncertainty of the Control System

The control system of the NanoPla has been optimised to reduce positioning errors. The remaining positioning errors are mainly caused by the resolution of the system components and electronic devices' noise. The computing operation in the control hardware is performed with finite numbers, which implies a rounding operation, resulting in a truncation error that, depending on its magnitude, may not be negligible. In addition, the errors due to electronic device noise cannot be completely eliminated and result in positioning noise. Thus, the positioning uncertainty of the control system is assessed and analysed in this section.

In Figure 13, the control system dataflow is represented in a block diagram. The data type of the transmitted information is represented by the coloured arrows. The control strategy is computed in Simulink® (MATLAB®). X_{ref} and Y_{ref} are the desired positions on the X- and Y-axes. The real position of the NanoPla is measured by a 2D laser system and extracted by MATLAB into the Simulink program. The control strategy computed in the PC by Simulink uses a 64-bit double-precision floating point format (blue arrows), and, in this case, the rounding operation has no significant influence on the calculated results. The control strategy outputs are the required phase voltages, contained in a range of ±6 V, and are sent to the control hardware by a serial communication interface (SCI) (green arrows). The MCU of the control hardware works with 32-bit data types, and the voltage values are transmitted as 32-bit fixed points with a 25-bit fraction length (red arrow). The resolution derived from the data type used for the voltages values is 0.0298 nV. These phase voltages are generated in the power stage of the control hardware by PWM. The DMC kit includes a high-resolution PWM (HRPWM) module that is capable of extending the time resolution capabilities of the PWM function. Thus, the resolution of the voltage generation is defined by the time resolution of the HRPWM module and is 26.1 µV. This resolution is sufficient to perform a minimum incremental motion of approximately 700 nm in an open-loop. Therefore, this hardware is not able to generate the exact combination of phase currents for every target position. Nevertheless, when working in a closed loop, the positioning controller is capable of partially correcting this error by switching between combinations of phase currents [14].

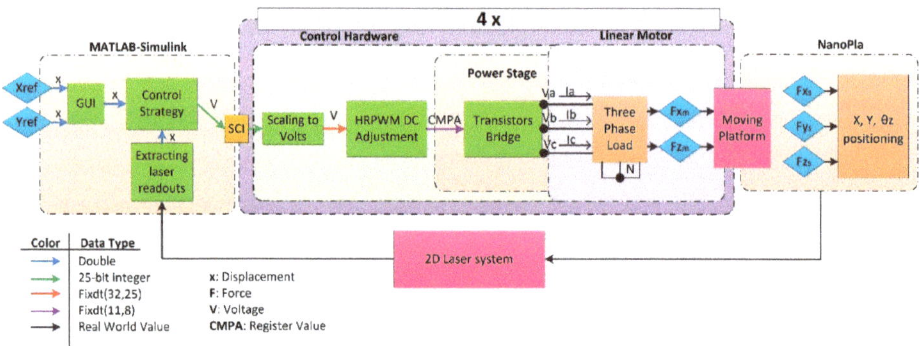

Figure 13. Block diagram of the data flow in the control system of the NanoPla.

In Figure 13, the real-world values are represented with black arrows. These values are the generated phase voltages, the derived currents, and the resultant forces. In addition, the 2D laser interferometer system measures the moving platform's displacement. The uncertainty of the laser system measurement also affects the positioning uncertainty of the control system.

It must also be taken into account that the PWM-controlled phase voltages lead to a ripple in the phase currents [15]. This current ripple is directly related to the inductance of the stator coils. The resistance and inductance of the coils have been experimentally measured and are 0.88 Ω and 0.24 mH in each phase coil. The electric circuit has been simulated in order to calculate the current ripple derived from the PWM-generated phase voltages. The current ripple has a sawtooth waveform with a frequency of 29.28 kHz, which is double the PWM frequency. Moreover, the current ripple peak-to-peak value is dependent on the duty cycle (DC) of the PWM voltages; in this study, the current's working range is ±0.83 A, which corresponds to a DC between 43.08% and 56.92%. For these values, the peak-to-peak value of the current ripple has a maximum magnitude of 0.09 A. The actual phase currents have been experimentally measured using a data acquisition system (DAQ) from National Instruments. This DAQ is able to record the measurements at a frequency of 500 kHz and with a resolution of approximately 2 mA. In Figure 14, an experimentally measured phase current is compared to the simulated one, and, as shown, the current ripple in both cases is almost coincident. Nevertheless, the experimentally measured phase current includes additional noise. This deviation has different sources, including the DAQ's own noise. One of the contributors is the noise of the DC power supply that feeds the power stage of the control hardware. The noise of the DC power supply is imprinted in the PWM phase voltages and, thus, is transmitted to the phase currents. To minimise this contributor, a low-noise power supply, with a peak-to-peak noise of 10 mV, has been used.

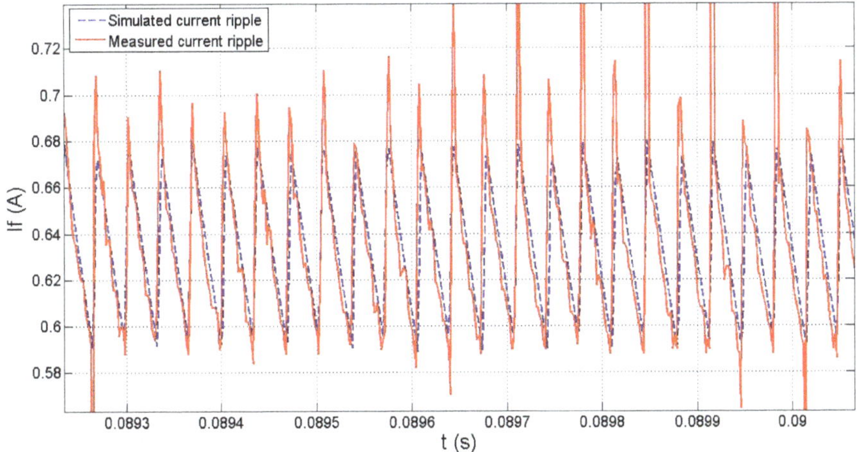

Figure 14. Current ripple of the phase currents generated by the pulse-width modulation (PWM)-controlled phase voltage.

In a previous study [16], a self-calibration procedure for the geometrical characterization of the 2D laser system of the NanoPla was proposed. The standard uncertainty of the calibrated laser system, after correcting for geometrical errors, was calculated to be 99 nm on the X- and Y-axes. The laser system's resolution is 1.58 nm, and the RMS deviation of the laser readouts of an axis is 6 nm. In addition, the stability of the 2D laser system integrated in the NanoPla was also verified.

In the control system of the NanoPla, the phase current noise generates deviations in the forces that act on the moving platform, thereby producing undesired vibrations in the platform. These vibrations are recorded by the laser system, adding to the laser system's own noise, and fed back to the control strategy. This results in positioning noise of the moving platform that has been experimentally measured and computed as a short term (30 s) RMS positioning error that is 0.11 μm in each axis. It has also been experimentally verified that the main contributor to the RMS positioning error is the phase current's noise.

NanoPla positioning uncertainty contributors can be divided into two categories: Ones whose contribution to the final positioning error in an open loop is known, like the resolution of the phase voltage generation (analysed in [14]) and the laser system (analysed in [16]); and the errors in the laser system that are still present after correcting for the geometrical errors obtained by the self-calibration procedure defined in [16]. The other types of errors are those whose contributions to the final positioning error cannot be calculated separately and cause RMS deviations of the positioning errors. Table 1 illustrates a calculation of the positioning uncertainty according to ISO/TR 230-9:2005 [17] and its contributors. Even though the laser system's resolution is included inside the standard uncertainty of the laser system, its value is also shown separately in the table, so it can be compared to the magnitude of the other contributors.

Table 1. NanoPla positioning uncertainty contributors and calculation.

Source	Justification	Standard Uncertainty	Relative Contribution
Resolution at the high-resolution PWM (PWM) u_{HRPWM}	Resolution of 26.2 µV	$0.7/\sqrt{12}$ µm	65.1%
Laser system resolution u_{Lres}	Resolution of 1.58 nm	$1.58/\sqrt{12}$ nm	0.00%
Laser system calibration u_{Lcal}	Geometrical errors + measuring system calibration [16]	99 nm	15.6%
RMS positioning error u_{RMS}	Laser system noise + phase currents noise + NanoPla vibrations	0.11 µm	19.3%
Positioning uncertainty $U_{XY}(k = 2)$	$U_{XY}(k = 2) = k\sqrt{u_{HRPWM}^2 + u_{Lcal}^2 + u_{Lres}^2 + u_{RMS}^2}$	±0.50 µm	100%

The resulting positioning uncertainty U_{XY} ($k = 2$) in each axis and in the entire working range of 50 mm × 50 mm is equal to ±0.50 µm. The main contributor is the HRPWM module resolution, which contributes partially to both. That is, when the laser system detects this error, the controller acts on the horizontal force to correct it, thereby resulting in oscillations. The phase currents' noise is another main contributor to the positioning uncertainty. None of these errors can be corrected without additional electronics. Nevertheless, the resulting positioning uncertainty is much lower than the initial working requirements of the NanoPla, so the developed positioning control system is considered valid.

5. Conclusions

In this work, a positioning control system for a 2D nanopositioning stage was designed and implemented in the NanoPla. The proposed control system drives four Halbach linear motors that allow planar motion, while a 2D plane mirror laser interferometer system works as the positioning sensor. The selected control hardware is a Digital Motor Control kit from Texas Instruments for generic three-phase motors. The target is to obtain an accurate positioning control system that fulfils the NanoPla requirements by implementing commercial hardware without any additional electronics.

The NanoPla offers a two-stage scheme that complements the XY-long range positioning of the moving platform (50 mm × 50 mm) with an additional commercial piezo-nanopositioning stage that is fixed to the inferior base. This second stage works in a range of 100 µm × 100 µm. Due to this, the maximum positioning error requirement of the NanoPla in the X- and Y-axes has been decided to be 10 µm. In addition, the rotation around the Z-axis must be kept minimal in order to avoid the laser system's misalignment. In this article, a dynamic model for a NanoPla with four Halbach linear motors as actuators was first identified. Then, a control strategy for the positions in the X- and Y-axes and the rotation around Z-axis was designed and implemented for the control hardware. The correct performance of the proposed control system has been experimentally verified for the NanoPla. In addition, the positioning uncertainty of the control system has been computed, and its contributors analysed. The obtained positioning uncertainty $U_X = U_Y = U_{XY}$ ($k = 2$) is equal to ±0.50 µm in each axis and in the entire working range of the NanoPla. Therefore, the resulting positioning uncertainty of the control system implemented in commercial generic hardware without additional electronics is much lower than the initial NanoPla requirements, thus broadening the applicability of the designed positioning system.

In future work, the control strategy could be improved to minimise the coupling between axes during movement. In addition, possible alternatives to improve global uncertainty with additional electronics will be studied. Future research should also focus on the implementation of a measuring system in the NanoPla. As previously noted, the target application of this first prototype is surface topography characterisation at the atomic scale of samples with relatively large planar areas, using an AFM. Nevertheless, due to the fragile configuration of the AFM system, the implementation of a confocal sensor as an intermediate solution before integrating the AFM is proposed.

Author Contributions: L.D.-P., M.T., J.A.A., and J.A.Y.-F. developed the positioning control system of the NanoPla.; L.D.-P. performed the experiments and wrote the original draft of the manuscript. M.T., J.A.A., and J.A.Y.-F. reviewed and edited the manuscript.

Appl. Sci. **2019**, *9*, 4860

Funding: Funded by the Gobierno de Aragón (Reference Group T56_17R) and co-funded with Feder 2014–2020 "Construyendo Europa desde Aragón" and the Spanish government project RTI2018-097191-B-I00 "MultiMet" with the collaboration of the Diputación General de Aragón—Fondo Social Europeo.

Conflicts of Interest: The authors declare no conflict of interest. The funders had no role in the design of the study; in the collection, analyses, or interpretation of data; in the writing of the manuscript, or in the decision to publish the results.

References

1. Gao, W.; Kim, S.; Bosse, H.; Haitjema, H.; Chen, Y.; Lu, X.; Knapp, W.; Weckenmann, A.; Estler, W.; Kunzmann, H. Measurement technologies for precision positioning. *CIRP Ann.* **2015**, *64*, 773–796. [CrossRef]
2. Sato, K. Trend of precision positioning technology. *ABCM Symp. Ser. Mechatron.* **2006**, *2*, 739–750.
3. Manske, E.; Jäger, G.; Hausotte, T.; Füßl, R. Recent developments and challenges of nanopositioning and nanomeasuring technology. *Meas. Sci. Technol.* **2012**, *23*, 74001–74010. [CrossRef]
4. Torralba, M.; Valenzuela, M.; Yagüe-Fabra, J.; Albajez, J.A.; Aguilar, J.; Fabra, J.A.Y. Large range nanopositioning stage design: A three-layer and two-stage platform. *Measurement* **2016**, *89*, 55–71. [CrossRef]
5. Trumper, D.; Kim, W.-J.; Williams, M. Design and analysis framework for linear permanent-magnet machines. *IEEE Trans. Ind. Appl.* **1996**, *32*, 371–379. [CrossRef]
6. Torralba, M.; Yagüe-Fabra, J.A.; Albajez, J.A.; Aguilar, J.J. Design Optimization for the Measurement Accuracy Improvement of a Large Range Nanopositioning Stage. *Sensors* **2016**, *16*, 84. [CrossRef] [PubMed]
7. Ruben, S. Modeling, Control, and Real-Time Optimization for a Nano-Precision System. Ph.D. Thesis, University of California, Los Angeles, CA, USA, 2010.
8. Hu, T. Design and Control of a 6-Degree-of-Freedom Levitated Positioner with High Precision. Ph.D. Thesis, Texas A&M University, College Station, TX, USA, 2005.
9. Diaz-Perez, L.C.; Torralba, M.; Albajez, J.A.; Yagüe-Fabra, J.A. One-dimensional control system for a linear motor of a two-dimensional nanopositioning stage using a commercial control hardware. *Micromachines* **2018**, *9*, 421. [CrossRef] [PubMed]
10. Diaz-Perez, L.C.; Torralba, M.; Albajez, J.A.; Yagüe-Fabra, J.A. Positioning uncertainty of the control system for the planar motion of a nanopositioning platform. Procedia Manufacturing. In Proceedings of the 8th Manufacturing Engineering Society International Conference, Madrid, Spain, 19–21 June 2019.
11. Lu, X.; Usman, I.-U.-R. 6D direct-drive technology for planar motion stages. *CIRP Ann.* **2012**, *61*, 359–362. [CrossRef]
12. Steinmetz, C. Sub-micron position measurement and control on precision machine tools with laser interferometry. *Precis. Eng.* **1990**, *12*, 12–24. [CrossRef]
13. Büchner, H.-J.; Jäger, G. A novel plane mirror interferometer without using corner cube reflectors. *Meas. Sci. Technol.* **2006**, *17*, 746–752. [CrossRef]
14. Díaz-Pérez, L.C.; Albajez, J.A.; Torralba, M.; Yagüe-Fabra, J.A. Vector Control Strategy for Halbach Linear Motor Implemented in a Commercial Control Hardware. *Electronics* **2018**, *7*, 232.
15. Jiang, D.; Wang, F. Current-ripple prediction for three-phase PWM converters. *IEEE Trans. Ind. Appl.* **2014**, *50*, 531–538. [CrossRef]
16. Torralba, M.; Díaz-Pérez, L.C.; Valenzuela, M.; Albajez, J.A.; Yagüe-Fabra, J.A. Geometrical Characterisation of a 2D Laser System and Calibration of a Cross-Grid Encoder by Means of a Self-Calibration Methodology. *Sensors* **2017**, *17*, 1992. [CrossRef] [PubMed]
17. International Organization for Standardization. *ISO/TR 230-9: Test Code for Machine Tools. Estimation of Measurement Uncertainty for Machine Tools Test according to Series ISO 230, Basic Equations*; International Organization for Standardization: Geneva, Switzerland, 2005.

Article

Hyperspectral Imaging Techniques for the Study, Conservation and Management of Rock Art

Vicente Bayarri [1,2,*], Miguel A. Sebastián [1] and Sergio Ripoll [3]

1. Manufacturing and Construction Engineering Department, ETS de Ingenieros Industriales, Universidad Nacional de Educación a Distancia, Calle Juan del Rosal 12, 28040 Madrid, Spain; msebastian@ind.uned.es
2. GIM Geomatics, S.L. C/Conde Torreanaz 8, 39300 Torrelavega, Spain
3. Paleolithic Studies Laboratory, Prehistory and Archaeology Department, Universidad Nacional de Educación a Distancia, Avda. Senda del Rey 7, 28040 Madrid, Spain; sripoll@geo.uned.es
* Correspondence: vicente.bayarri@gim-geomatics.com; Tel.: +34-635-500-584

Received: 20 October 2019; Accepted: 19 November 2019; Published: 21 November 2019

Featured Application: This work presents several hyperspectral techniques that can provide a scientific and non-destructive method for the study, conservation and management of rock art. Tasks such as recognition of coloring matter, formal recognition of the figures, superposition of forms and documentation of the state of conservation can be solved efficiently with hyperspectral imaging.

Abstract: Paleolithic rock art is one of the most important cultural phenomena in the history of mankind. It was made by making incisions and/or applying natural pigments mixed with water or organic elements on a rock surface, which for millennia has been subjected to different factors of natural and anthropogenic alteration that have caused its deterioration and/or disappearance. The present paper shows a methodology that employs hyperspectral technology in the range of visible light and the near infrared spectrum, providing a scientific and non-destructive way to study, conserve and manage such a valuable cultural heritage. Recognition of coloring matter, formal recognition of the figures, superposition of forms and documentation of the state of conservation are relevant topics in rock art, and hyperspectral imaging technology is an efficient way to study them. The aim is to establish a method of creating pigment cartography and enhancing the visualization of rock art panels. Illumination sources, spectroradiometry measurements and camera adjustments must be taken into account to generate accurate results that later will be pre-processed to derive reflectance data, and then pigment analysis and enhanced visualization methods are applied. This methodology has allowed us to obtain 76% more figures than using traditional techniques throughout the case study area.

Keywords: hyperspectral imaging; pigment analysis; cultural heritage; conservation; cultural management; visual enhancement; rock art; cartography

---

1. Introduction

Prehistoric art is an extraordinary manifestation that documents, with great detail and precision, some facts and animal species that coexisted with the humans of Paleolithic times. It is the first industrial and technical heritage of history, and such a singular phenomenon must be protected [1]. It was made by making incisions and/or applying natural pigments mixed with water or organic elements on a rock surface, which for millennia has been subjected to different factors of natural and anthropogenic alteration that may have caused its deterioration or disappearance.

The deteriorating factors can act in an isolated or combined way on this cultural heritage and its support, and include dirt, impacts, flaking, disintegration, washing, glazes, coatings, etc.

The conservation of industrial, technical and cultural heritage requires a deep understanding of the significance and complexity of a place. Good conservation of our heritage is based on informed decisions, and good documentation ensures that knowledge of heritage places will be passed on to future generations.

This work analyses, develops and implements a methodology to generate information not visible to the naked eye due to dirt, washing, glazes and coatings. This information not only improves the visualization of the panels, but also helps us to understand how they were created and sheds light on the stratigraphy or sequence of the execution of different motifs.

The methodology is applicable to any rock art panels, and any cultural or industrial heritage objects in general, that need to be studied or controlled in time. The method has been tested on a panel that has been studied since the beginning of the 20th century, which presents a series of complex cases to solve such as superposition of paints, calcite glazes, pigment–sketch combination, etc.

Traditional rock art documentation systems are subjective [2], as an individual decides, usually by direct visualization, whether a unit is part of the panel or not. Throughout history, harmful methods such as casts and direct molds have been used, as well as others such as analogue, digital, ultraviolet and infrared photography, which have contributed to create "digital tracings" from images processed using image software. In all of them, it is an operator who decides whether or not the panel belongs to the class. Such subjectivity must be avoided, and hyperspectral remote sensing is an optimal tool for this.

A plethora of techniques, analyses and applications for hyperspectral remote sensing have been developed in the last few decades [3]. It has been used for the documentation of murals [4], detection of buried archaeological relics [5], and exploration of their geologic [6] or mineralogic composition. It has a number of advantages [7]: It is considered a safe detection technique because it is non-contact and non-destructive. Another advantage is its high spectral resolution with multiple bands and narrow bandwidth, i.e., it allows a continuous spectral response to be obtained for each pixel of the image. Because of these unique features, hyperspectral images have been used in conservation [8] and the study of artefacts [9] and to reveal hidden details in ancient manuscripts and paintings, although its use in rock art [10], and in cultural and industrial heritage objects [11], has been scarce and relatively recent.

The main objective of this study is to propose a method of generating the cartography of pigments and improving the visualization of aspects of rock art not visible to the naked eye due to the causes previously mentioned, as well as to understand its process of creation or the sequence of execution of different motifs.

2. Materials and Methods

2.1. Study Area

The study area corresponds to a series of panels in the Cave of El Castillo in Puente Viesgo, Cantabria, Spain, which was declared a World Heritage Site by United Nations Educational, Scientific and Cultural Organization (UNESCO) in 2008.

Today this cave has the world's oldest Paleolithic art, at least 40,800 years old. The cave contains animal figures such as bison, horses, deer, mammoth, aurochs, goats, etc. The panels also feature hands, dots and rectangular shapes. Also found is one of the oldest industrial and technical heritage sites, associated with bone and lithic industry.

It is possible to find engravings with different characteristics and paintings that have red, black or yellow colors, applied with finger, airbrush, pencil or brush. The state of conservation of the panels is not very good. The panels have suffered graffiti, glazes, cleanings, etc., that make the motives unrecognizable with the naked eye. In order to reveal latent information, hyperspectral data from the panels have been captured and analyzed.

The panel de las manos of the cave of El Castillo (Figure 1) is possibly one of the most complex among Paleolithic art in general. The use of different techniques and colorants allows us to establish differences between groups and help us understand the chronostylistic evolution based on the superpositions of figures. The chronological frame of the figures has been set between 41,400 and 10,000 years before the present (BP) thanks to uranium-thorium and radiocarbon dating, although we have focused on the overlays of the Gravettian-Aurignacian phases. The panels were first studied by Alcalde del Río [12] in 1911, but during the last century many researchers have revisited the site and have applied different techniques to study the rock art panels of the cave such as hand-drawing, direct tracing, rubbing or *frottage* tracing; making molds directly from the panels; analogue photography; digital images; and techniques to enhance visualization such as decorrelation of individual elements [13–18].

Figure 1. (**a**) View of the 3D model of panel de las manos. (**b**) Orthoimage generated to register all hyperspectral data.

As will be shown, hyperspectral remote sensing can help not only to better identify figures, but also isolate them by type of dye and analyze superimpositions.

In the first study carried out on the panel de las manos [12], 9 animals painted in yellow, 33 negative hands, 13 painted signs and some groups of red punctuation, and 6 engraved figures, mainly deer, were identified. Using hyperspectral technology, 35 zoomorphs (23 in yellow, 2 in red, 19 in black and 1 in white), 56 hands (in 2 phases) and 22 signs have been documented today [18].

The study was carried out on the panel as a whole; the archaeological interpretation was made by Ripoll [19]. Nevertheless, the present study shows three areas of special interest and the hand analysis of the whole panel.

2.2. Overall Workflow

Figure 2 shows the overall workflow of the proposed method for the data acquisition, pre-processing, pigment analysis and extraction of the hidden information of rock art. The details of each step are discussed in the following subsections.

The data acquisition was carried out in 2014 and 2018. Hyperspectral imaging data were acquired with a VNIR Specim V10E sensor and pre-processing steps were applied to transform raw data from radiance into reflectance values and were georeferenced and geometrically corrected.

Pre-processed hyperspectral images were classified using spectral angle mapper (SAM) and mixture tuned match filtering (MTMF) [20] image processing algorithms to generate a pigment cartography. End members for classification were selected from pure samples identification spectral analysis and were sampled by using a field spectroradiometer. In parallel, a process to enhance visualization was used in order to generate false color compositions to ease the pigment interpretation.

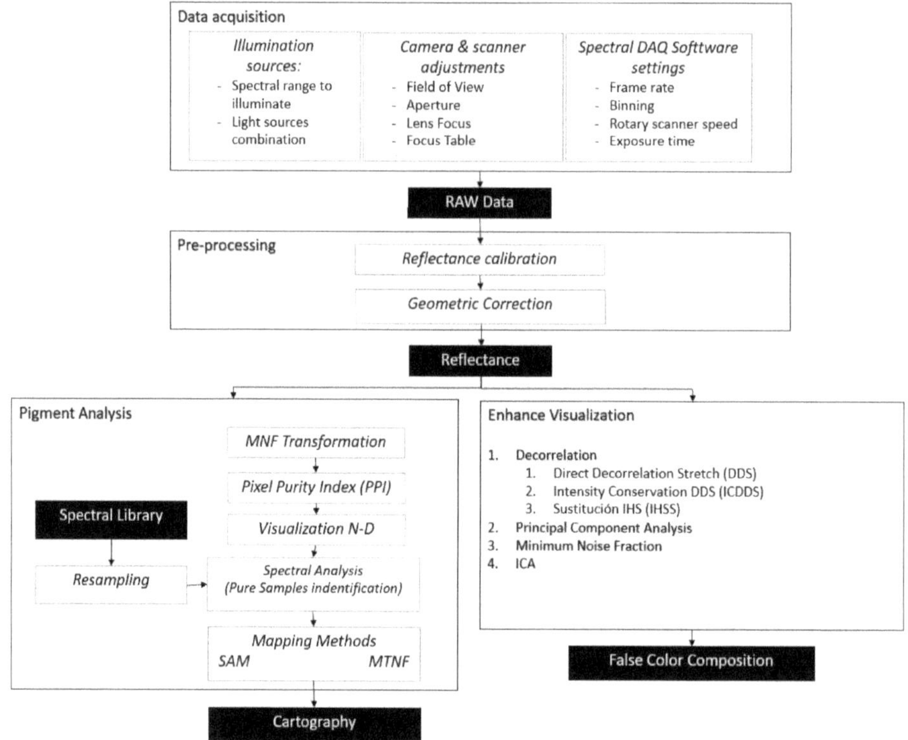

Figure 2. Overall workflow of the proposed method.

2.3. Data Acquisition

2.3.1. Hyperspectral Imaging System

The hyperspectral imaging system used in this study consists of a Specim V10E spectrograph (Spectral Imaging Ltd., Oulu, Finland) covering a spectral range of 400–1000 nm, a sCMOS camera (CL-30, effective resolution 1312×728 pixels by 12 bits), an objective lens (Cinegon 2.4/30 mm focal length), a rotating scanner light and data acquisition (DAQ) spectral software.

The 2.8 nm spectral resolution was reduced to 5.6 nm by applying binning to improve the signal-to-noise ratio. The data were recorded as a hypercube with two-dimensional spatial images and wavelength bands as the third dimension, and 214 spectral bands were extracted from each pixel to form a relatively continuous spectral curve.

2.3.2. Illumination Sources

The cave of El Castillo is a place with very stable conditions, with an average temperature of 13.8 °C, humidity of 99% and total absence of light.

This makes it necessary to illuminate the panels with artificial light, for which the recommendations of the International Council of Museums (ICOM) were followed [21]: section 9, "physical environment", recommends that moderately sensitive elements should only be exposed to filtered ultraviolet light with an intensity not exceeding 150 lux. This was necessary to create a lighting setup for the field campaigns in 2014–2018, which was calibrated in the spectral range where the sensors were operational.

For this purpose, ultraviolet and infrared light-emitting diode (LED) lights and tube lights were used. The lighting setup was formed by an array of four Phillips TL5 tubular fluorescent lamps (green), ultraviolet and infrared LED lamps. The spectral signature of the lights was obtained from the 1 nm wide narrowband FieldSpec Pro FR spectroradiometer manufactured by Analytical Spectral Devices measuring spectra over a range of 350–1050 nm (Figure 3).

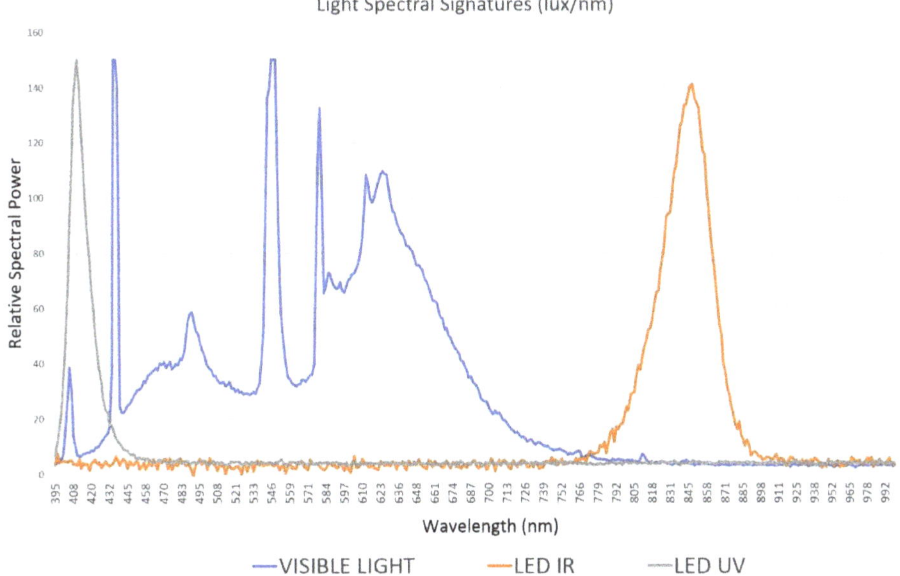

Figure 3. Spectral signatures of lights at working distance. Vertical axis represents relative spectral power measured and horizontal axis the wavelength in nanometers where it was measured.

2.4. Pre-Processing

2.4.1. Reflectance Calibration

Prior to image acquisition, a protocol to eliminate the effects of black (B) and white (W) background on the hyperspectral images was followed. A white image was acquired by recording a uniform white target, and a black image by closing the shutter of the camera. The calibrated reflectance values (R) of raw sample images (I) were calculated by the following formula:

$$R_i = \frac{I_i - B_i}{W_i - B_i}$$

where i is the pixel index.

2.4.2. Geometric Correction

A hyperspectral image is a conical projection of reality. The geometric correction process is carried out to eliminate errors or distortions that occur in the image, and to adapt the image to an established 3D reference system that allows it to be compared.

The steps to follow in the geometric correction of an image are as follows:

1. Select control points.
2. Calculate the transformation.
3. Resample the image.
4. Verify the process.

In this case, the correction was made by projecting the hyperspectral image onto the 3D model of the panel de las manos obtained by photogrammetry, which was subsequently re-projected following an orthogonal plane (Figure 1). The hyperspectral imaging camera had certain limitations regarding the imaging range and resolution. The pixel size was set to 1 mm and each recording had approximately 1 m^2.

2.5. Pigment Analysis

Integrated processes are classified into two families:

- Analysis techniques, where the aim is to be able to generate pigment cartography and distinguish between those that are equal and those that are not.
- Enhance visualization techniques, where the aim is to be able to recover paintings that cannot be seen with the naked eye.

The result of the pigment analysis is thematic cartography, and each pixel has an assigned class value, but the result of the enhanced visualization technique is images.

2.5.1. Minimum Noise Fraction Transformation

The modified minimum noise fraction (MNF) transformation was implemented as described by Green [22]. It is a technique that transforms the data hypercube into a set of canals with increasing noise levels by means of noise whitening and the subsequent application of main components. The MNF transformation is also used to eliminate noise from data by performing a direct transformation, determining which bands contain coherent images (examining images and auto-values), and then executing an inverse transformation using a spectral subset that includes only good bands, or smoothing the noisier bands before executing the inverse.

2.5.2. Pixel Purity Index

The pixel purity index (PPI) is used to find the most spectrally pure (extreme) pixels in multispectral and hyperspectral images. In our case it was used to distinguish between different types of ochre and black pigments. It normally corresponds to the final mixing members. The PPI is calculated by repeatedly projecting n-D scatter diagrams on a random unit vector.

Computationally, extreme pixels are recorded at each projection (pixels that fall on the ends of the unit vector), and the number of times each pixel is marked as extreme is recorded. A pixel purity image is created where each pixel value corresponds to the number of times the pixel was recorded as extreme. The PPI function can create a new output band or continue its iterations and add the results to an existing output band. The PPI is typically executed in an MNF transformation result, excluding noise bands. PPI results are generally used as input in the n-D viewer.

2.5.3. n-D Display

The n-D Visualizer is an interactive tool in the ENVI system that is used to locate, identify and group the purest pixels and the most extreme spectral responses in a dataset. The n-D viewer helps to visualize the shape of a data cloud that results from plotting image data in the spectral space (with image bands as frame axes). A spatial subset of MNF data using only the purest pixels determined from the PPI is normally used. It is also used to check the separability of classes when generating regions of interest (ROIs) as an entry in supervised classifications.

2.5.4. Spectral Angle Mapper

The spectral angle mapper (SAM) was developed by Boardman [20], but we based ours on Rashmi [23]. It is a physics-based spectral classification that uses an n-D angle to match pixels to reference spectra. The algorithm determines the spectral similarity between two spectra by calculating the angle between the spectra and treating them as vectors in a space with dimensionality equal to the number of bands.

This technique, when used with calibrated reflectance data, as in the present case study, is relatively insensitive to the effects of lighting and albedo. The spectra of pure samples used by SAM were recorded using the Analytical Spectral Devices full-resolution spectroradiometer, as described in Section 2.5.5. SAM compares the angles between final member spectrum vectors and each pixel vector in n-D space. Smaller angles represent coincidences closer to the reference spectrum. Pixels further than the maximum angle threshold specified in radians are not classified.

2.5.5. Field Spectroradiometer

Spectral reflectance signatures usually help to determine whether two pigment types can be distinguished from each other, and if so, in which part of the spectrum the spectral characteristics differ.

The sampling was made by using the Analytical Spectral Devices full-resolution spectroradiometer. This spectroradiometer can obtain data in the range of 350–2500 nm (Figure 4), but it's important to note that the hyperspectral imaging system will take advantage of data only between 400 and 1000 nm. Spectral signatures of the pigments were obtained, and it was especially important to characterize the places where the spectral analysis detected the end members or pure samples of ochre pigment in order to train the later process of mapping methods.

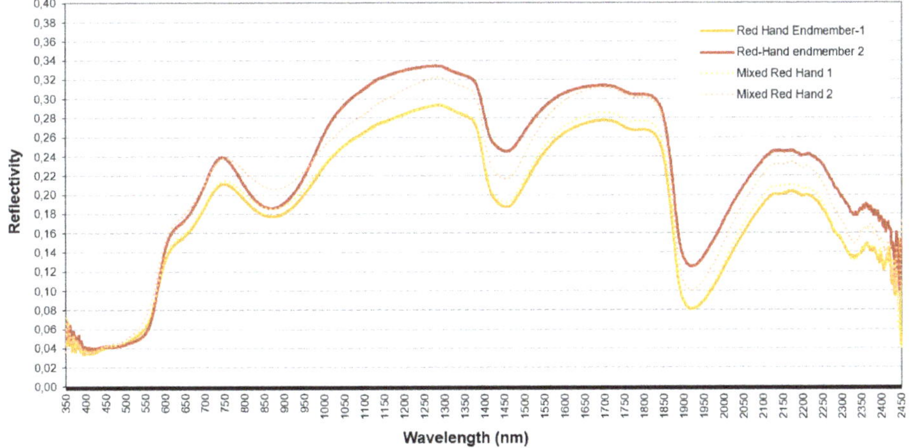

Figure 4. Spectral signatures of ochre pigment in hands.

With the data, it is possible to differentiate between different types of pigments that to an observer's eye would be almost indistinguishable. This is possible because the spectral signature, instead of using information from the visible spectrum, integrates parts of the ultraviolet and infrared spectra, where the different composition offers different signals. Spectral signature data are used to analyze the separability between classes, that is, to know how many types of pigment can be distinguished and then integrated into the processing of clouds.

2.6. Visualization Improvement

2.6.1. Decorrelation Adjustment

Decorrelation adjustment is implemented for generic binary band data stored in 16 bits, as described by Sabins [24], and is used to enhance the image. It consists of converting the data to the space defined by the main components of the original bands, followed by equalizing the data according to the new axes, and finally converting the data to the initial space and combining the bands with the primary colors, red/green/blue (RGB). In this way, the dots are more evenly distributed in the RGB space, so the image will show much higher contrast.

2.6.2. Principal Component Analysis

Two principal component algorithms were used, one adapted from Richards [25] and the Karhunen–Loeve transformation, as described in Loeve [26], which were programmed in the IDL language. Both methods were carried out using correlation matrices and all bands were included in the transformation.

2.6.3. Minimum Noise Fraction Transformation

The modified MNF transformation was implemented as described by Green [22].

2.6.4. Independent Component Analysis

Independent component analysis (ICA) can be used in multispectral datasets to transform a set of randomly mixed signals into mutually independent components according to Hyvarinen [27,28].

3. Results and Discussion

The panel de las manos of the cave of El Castillo is possibly one of the most complex among Paleolithic art in general. The use of different techniques and colorants allowed us to establish differences between groups and help us understand the chronostylistic evolution based on superpositions of figures. It has been studied repeatedly by many researchers since 1911 [13–18].

This study was carried out on the panel as a whole; the archaeological interpretation was made by Ripoll [19]. The present study shows three areas of special interest and the hand analysis of the whole panel (Figure 5).

Figure 5. Locations of four areas of special interest in the study.

3.1. Working Area 1

This area consists of a red bison figure, made of iron oxide with black graphite remains of old graffiti, which was cleaned in the 1960s (Figure 6).

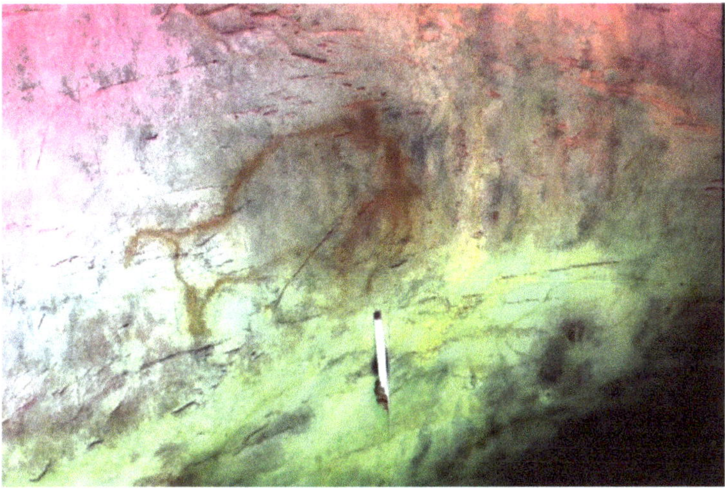

Figure 6. True color composition of hyperspectral image.

After applying the workflow, the following can be concluded:

1. Thanks to enhanced visualization algorithms, it was possible to recover the processes of cleaning of superpositions and observe graffiti that had been previously eliminated (Figure 7), so that visitors can have a more real perception of the prehistoric images. In the same way, this information allowed us to document the surfaces of the Paleolithic images that were affected by cleaning. This appeared in both MNF and ICA transforms.

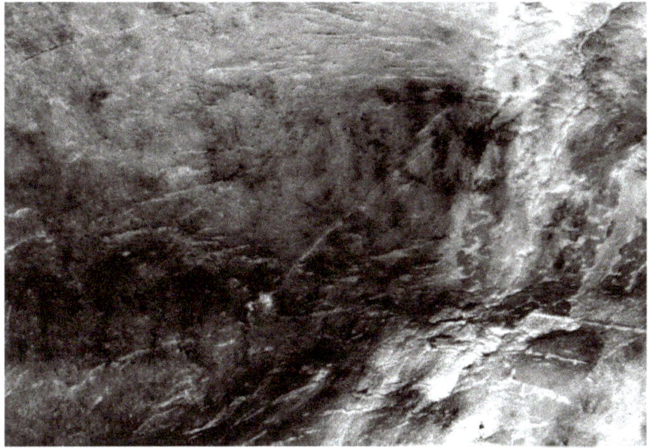

Figure 7. Remains of pigments extracted in the minimum noise fraction (MNF)-6 component.

2. Spectral differentiation according to the different mineralogical compositions of the pigments was possible, possibly associated with repainting. Figure 8a shows ochre remains of a bison with the

same spectral signature as the lower left hand side, while the left image signal shows the same ochre as one of the "yellow" phases between the two hand phases. This allowed us to say that the bison was painted simultaneously with the first hands (Figure 8a) and later repainted with other figures (Figure 8b).

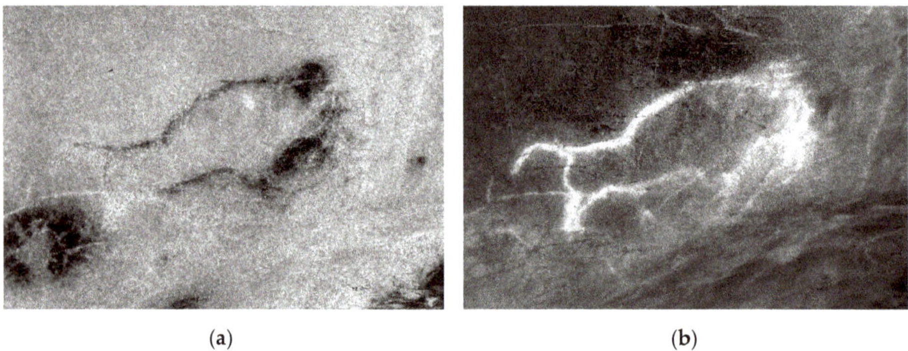

(a)   (b)

Figure 8. Cartography of (**a**) redder ochre type and (**b**) a more yellowish one.

3. When pigment types were isolated, it was possible to recover the pigments; the limits of the contours of the figure were clearly documented, and a formal reconstruction of the motif was carried out depending on the type of pigment, as shown in Figure 9.

Figure 9. Spectral angle mapper (SAM) classification of pigments.

3.2. Working Area 2

In this area, two elements were included (Figure 10):

1. A rectangular geometric pattern: a yellow drawing of flat ink using iron oxide. The drawn motif was contoured, exceeding the contours of the drawing by means of repeated external contour engraving.

2. A complex geometric and linear motif formed by two slightly concave rectangular structures and finished off at the top by small strokes associated with continuous lines. Probably prior to the drawing of the motif, a light layer of red dye was applied to the substrate. The drawn motif was contoured, surpassing and superimposing by means of contour engraving repeated externally to the contours of the drawing.

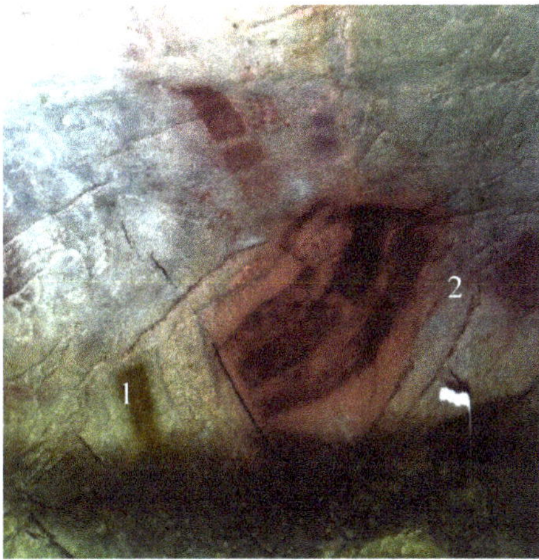

Figure 10. True color composition of hyperspectral image showing motifs: (**1**) is the rectangular geometric pattern and (**2**) is the complex geometric and linear motif.

3.2.1. Element 1

1. A reconstruction of the technical process was possible. The reading of the image allowed us to study in detail the traceability of the lines that compose the figure, which allowed us to approach the analytical reconstruction of the superposition of lines within the same line (Figure 11a).

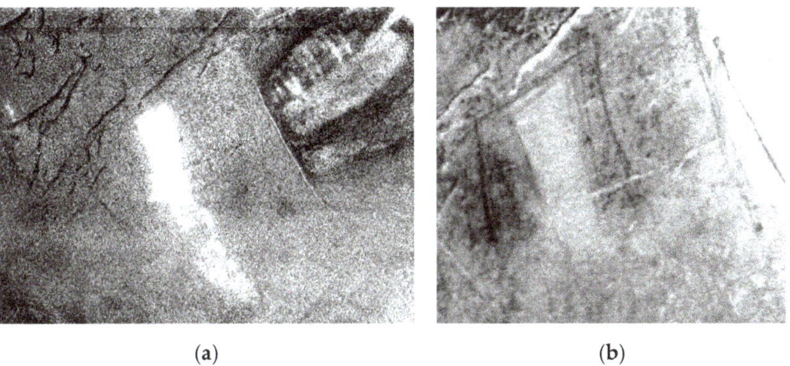

(a)   (b)

Figure 11. (a) Pigment represented in white; (b) engravings around the pigment.

2. Reconstruction with full sharpness of the delineation of engraved strokes and consequently of the engraving was possible (Figure 11b).

3. Recovery of the pigments: the limits of the contours of the figure were clearly documented and a formal reconstruction of the motif was carried out.

Study of the execution process: superpositions between figures were observed, which allowed reconstruction of the stratigraphy and the process of adding motifs (Figure 12).

Figure 12. False color composition showing pigments and engravings.

3.2.2. Element 2

4. Reconstruction of the technical process: the reading of the image made it possible to identify the possible existence of a layer of "plaster" (Figure 13a) prior to the drawing of the rectangular shapes, which made it possible to approach the graphic process.

| (a) | (b) |

Figure 13. (**a**) Plaster layer represented in black; (**b**) pigment drawing represented in white.

5. Reconstruction with full sharpness of the delineation of engraved strokes and consequently of the engraving was possible.

6. Study of the execution process: in both images superpositions between figures were observed, which allowed reconstruction of the stratigraphy and the process of adding motifs.

7. Precise formal identification: the limits of the contours of the figure were clearly documented and a formal reconstruction of the motif was carried out (Figure 13b).

8. Isolation of spectrally different pigments: Figure 14a shows (in black) coloring matter with different mineralogical composition that belongs to element 1 inside the motif. It was detected that the same pigments used in element 1 were used for eight strokes in element 2 of the panel, which were hardly perceptible to the naked eye (Figure 14b).

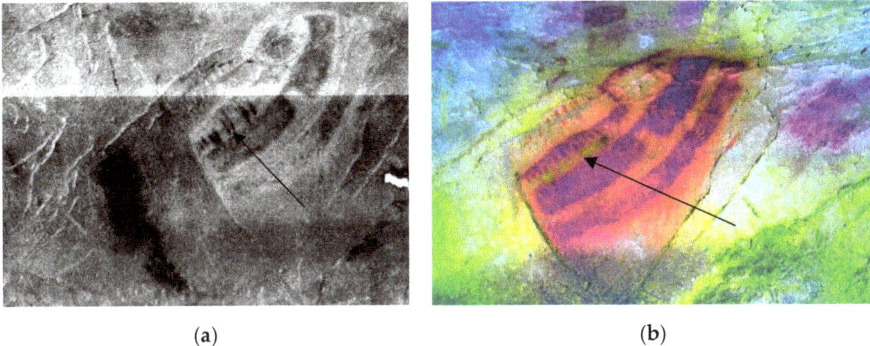

(**a**)   (**b**)

Figure 14. (**a**) Isolation of pigment of part 1 of the panel; (**b**) eight strokes (in green) made with pigment of part 1 in part 2.

3.3. Working Area 3

In this panel (Figure 15), above the back of the horse (point 3 in the figure) there is a small figure of a red female deer (1), incomplete and oriented to the right like the horse, with whose back it coincides. There is also a small horse infraposed (2) with the most evident yellow equine. It has a smaller size and is oriented to the right, although the anterior part practically coincides with the infraposition. On the left side there is a single tectiform in the shape of a slightly curved rectangle (4). The support has a very low degree of humidity, and the front part of the animal (especially the head) has a calcite coating. There is a yellow drawing of continuous tracing by means of iron oxide.

Figure 15. True color composition of hyperspectral image: (**1**) is the red female deer, (**2**) small horse infraposed, (**3**) visible horse and (**4**) is a single tectiform.

After applying the workflow, the following was concluded:

1. Areas affected by calcite coating due to lower signal strength were identified.
2. Recovery of the pigments: the limits of the contours of the figure were clearly documented and a formal reconstruction of the motif was carried out.
3. Study of the execution process: in both images, superpositions between figures were observed, which allowed reconstruction of the stratigraphy and the process of adding motifs.
4. Reconstruction of the technical process: reading of the image allowed us to study in detail the traceability of the lines that compose the figure, allowing us to approach the analytical reconstruction of the superposition of lines (Figure 16a,b) within the same line and study the means of applying the coloring matter.

5. Pigment discrimination: as a result of the SAM classifier, spectrally different areas were documented, which were clearly related to coloring matter of different mineralogical composition.

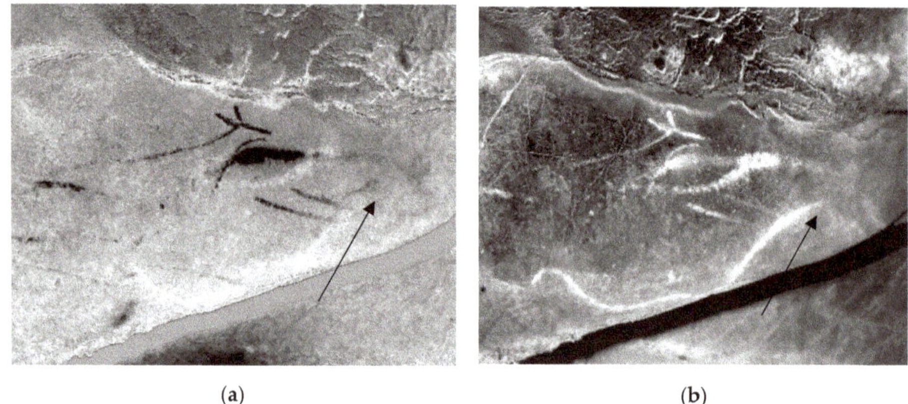

(a)   (b)

Figure 16. (**a**) As shown in black, it is possible to recover a horse below the calcite and yellow horse; (**b**) horse between calcite and old horse.

3.4. Working Area 4

Traditionally, silhouettes of hands are always at the base of pictorial stratigraphs. Those with absolute dates are located between the Gravettian and the beginning of the Solutrean periods. One of the traditional discussions on the panel of the hands is the position of the hands on the stratigraphy, that is, what layers they are above and below.

A great contribution of hyperspectral technology was used to differentiate and isolate two phases of hands based on their spectral signatures. Once isolated, by superimpositions it was seen that there are three pictorial layers between them (Figure 17a–d).

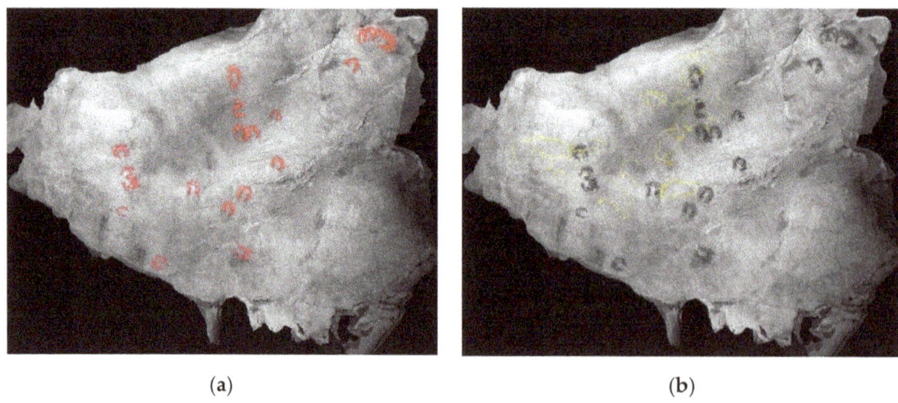

(a)   (b)

Figure 17. *Cont.*

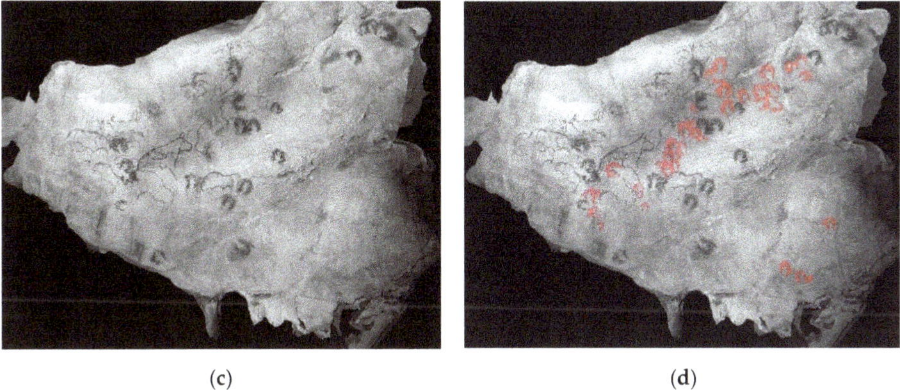

(c)   (d)

Figure 17. Images corresponding to (**a**) the first phase of hands (in red); (**b**) the second phase, elaborated with yellow pigment or sienna, very distinguished and centered in the set of aurochs; (**c**) the third phase with zoomorphic representations and two quadrangular signs or tectiforms painted in yellow pigment (shown in black); and (**d**) the fourth phase, which corresponds to the second phase of hands.

4. Discussion

4.1. Working Area 1

This panel has been studied several times throughout history because it is one of the best known of the world's rock art. If we compare the new results (Figure 18b) with the first publication in 1911 (Figure 18a), we can appreciate the following.

Ruiz-Redondo [18] affirms that all the animals were made in sienna, except for one black and one orange bison, and several of them are superimposed on negative hands. Ripoll [19] reported, based on hyperspectral remote sensing results, that there are two phases of hands and two phases of animals (first in ochre and second in black) between them.

The analysis derived from hyperspectral analysis data for the same area shows not only that it was capable of separating pigments, but also that new figures were emerging.

What has been traditionally represented is a red bison with the back well drawn, but the front part imperfect and confused. Thanks to the techniques employed, it was possible to clearly document the boundaries of the contours of the figure and carry out a formal reconstruction. Spectral differentiation was also performed according to the different mineralogical compositions and the identification of areas affected by leaching and calcite coating due to lower signal intensity.

Below the bison, the silhouette of another one facing to the left can be seen. Crossing the rear extremities of the bison (in yellow) and superimposed on the hand, a bovine silhouette can be seen, arranged to the left and somewhat wild. In addition to the above, new figures such as a horse's head appear in the central part of the bison.

4.2. Working Area 2

The drawing made in 1911 by Alcalde del Río [12] is shown in Figure 19a and the results of the SAM classification in Figure 19b, showing in yellow the pigments of element 1 that were used to make some of the strokes of element 2. In element 1 and part of element 2 we also saw that the whole contour was engraved with a very characteristic multiple stroke (in black).

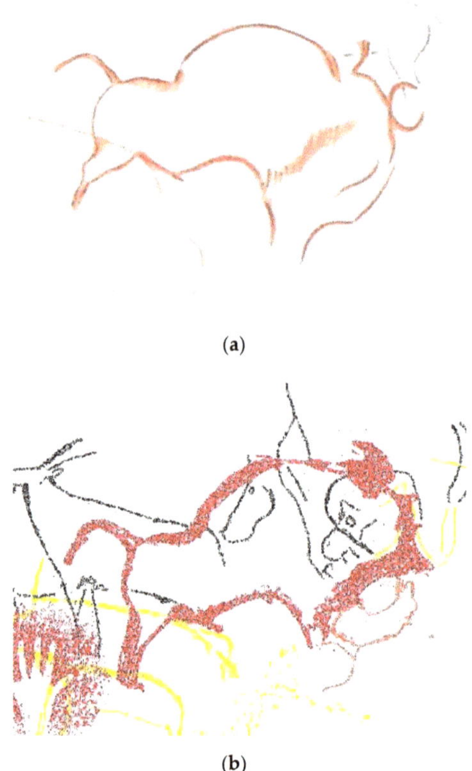

(**a**)

(**b**)

Figure 18. (**a**) Interpretation made in 1911 by Alcalde del Río [12]; (**b**) fragment of the SAM classification.

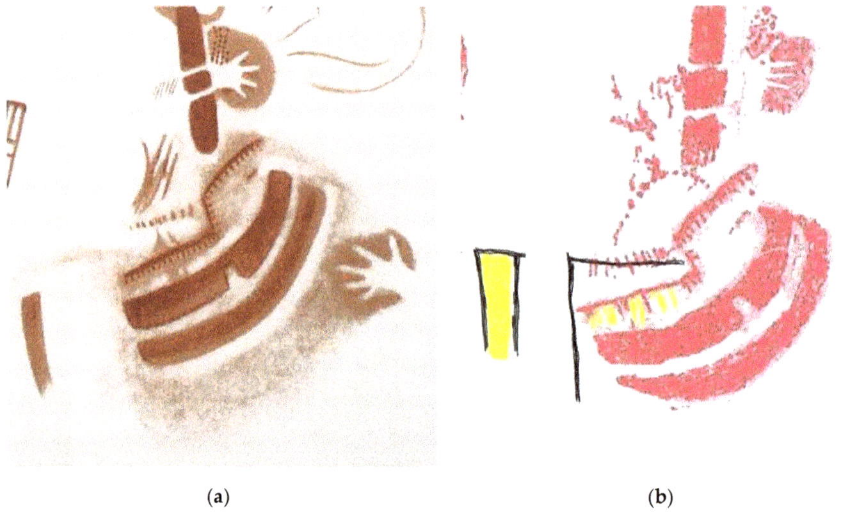

(**a**) (**b**)

Figure 19. (**a**) Interpretation made in 1911 by Alcalde del Río [12]. (**b**) Eight strokes (in green) made with pigment of part 1 in part 2.

4.3. Working Area 3

This panel also has been studied several times. Comparing the new results with the first publication in 1911 [12] (Figure 20), we found the following.

| (a) | (b) |

Figure 20. (a) Interpretation made in 1911 by Alcalde del Río [12]. (b) Detail of the red female deer in [12].

Garate [17] affirms that the horse is represented in a partial way, including the line of the corrected mane and the start of the back, the parallel ears straight, the nose opened by two lines also parallel and with the same stroke in the mouth as in the previous case. It also has the line of the chest and the ventral line with a marked sinuosity, although apparently the rear train was not included.

The results of the SAM classification (Figure 21) completed the figures to a great extent and showed everything described above.

- Figure 1: Above the back of the horse is a small figure of red female deer, incomplete and oriented to the right like this horse, with whose back it coincides.
- Figure 2: There is a small horse superimposed on the most obvious yellow equine. It has a smaller size and is oriented to the right, although the anterior part practically coincides with the infraposition. Thanks to hyperspectral analysis, we were able to identify this new figure, which was not seen directly due to the existence of a calcite veil. Recovery of the pigments allowed the contours of the representation to be clearly documented. With this methodology we observed the superpositions and reconstructed the iconographic stratigraphy and the process of addition of the motifs. We also discriminated the different pigments that are documented in spectrally different areas, which can be clearly related to coloring matter with different mineralogical compositions.
- Figure 3: There is another horse of ochre color, arranged to the right with a large belly and without legs. Its ears are upside down. It is above the tectiform. On the left, there is a red stroke and other blacks that have vanished.
- Figure 4: In the area of the rump of the horse, a simple tectiform in the shape of a slightly curved rectangle appears.

Figure 21. Result of SAM classification of work zone 4.

The results can be overlaid on the orthoimage since the data are orthocorrected and georeferenced (Figure 22).

Figure 22. Superposition of SAM classification over orthoimage of the panel in working area 3.

5. General Results

Working zones 1, 2 and 3 are part of a larger element known as the panel de las manos in the cave of El Castillo. The integration of techniques such as photogrammetry and remote sensing made it possible to obtain 76% more figures throughout the panel, mainly because they presented problems such as those previously described. The full panel review offered the results shown in Table 1 and graphically shown in Figure 23.

Table 1. Summary of results in full panel de la manos review.

Figures	Alcalde del Río et al. [12]	Hyperspectral Review
Hands	33	56
Bisons	9	17
Aurochs		3
Cervids		3
Caprids		3
Ideomorphs		11
Engravings	6	11
Total	7	97

Comparison of types of figures detected.

(a)

(b)

Figure 23. (a) Interpretation made in 1911 by Alcalde del Río [12]. (b) New pigment cartography obtained from hyperspectral data of the panel de las manos.

6. Conclusions

Methodology such as that presented for the documentation of rock art is essential in order to acquire knowledge to advance the understanding of cultural heritage and its evolution, and to encourage the interest and participation of people in heritage preservation. The revision of the three panels has resulted in a considerable increase in the number of known figures. In the panel de las manos, we see an increase from 55 to 97 elements (+76%).

Today, recording methodologies should use non-intrusive techniques and should not cause damage to the item being recorded; furthermore, before any data recording, the method to be used in the documentation should be clearly established.

The integration of geomatic methods allows documentation of any industrial or technical heritage in a non-intrusive, rigorous, safe and detailed way. The combined use of photogrammetric techniques supported on a rigorous topographic base makes the documentation of rock art panels more sustainable, as it allows us to reduce the time spent in the cave, which reduces the impact on it; avoid the use of target elements, which leads to better conservation of the cave; increase the accuracy of the support points: 2 mm of one point compared to 5–7 mm of the prismless measuring systems of total topographic stations; have control points throughout the panel to analyze deviations from the photogrammetric model.

Furthermore, these techniques are capable of delivering 2D and 3D results that can be used in any studio.

With the present case study, we sought to cover four areas of work that had high potential in the study of this type of graphic manifestation. The areas of interest were: (a) recognition of the coloring matter; (b) formal recognition of the figures; (c) recognition of the technical process; and (d) documentation of the state of conservation.

Once the work zones have been processed, we can conclude that in recognition of the coloring matter. Most of the techniques applied to the study of the composition of the coloring matter imply taking samples, which entails extractive action and, therefore, deterioration of the cave. Reading images worked at different spectral amplitudes makes it possible to clearly differentiate different compositions from the coloring materials used. This is an important milestone on which further work is undoubtedly needed, as compositions have so far been relatively differentiated, although further experiments with natural pigments will be necessary in order to obtain precise patterns and be able to link certain spectra with precise mineralogical compositions. The development and deepening of this field will make an important contribution, in the medium or long term, to the knowledge, at least relative, of the composition of the pigments without taking samples.

Reconstruction of the cave motifs. Most figures located inside cavities are subject to natural and artificial processes that cause the loss of coloring matter or the erosion of engraved surfaces, implying difficult interpretation of the motifs. The application of the technique has made it possible to precisely define the original morphology of some figures, both engraved and painted in different colors (and very probably of different chemical compositions). In particular, it is possible to precisely define the contours of the figures and precisely recognize anatomical parts or areas of specific figures, and consequently obtain images that involve highly reliable reconstruction of the original painting or engraving. In this way, figures that currently present visualization difficulties can be "reconstructed" and thus allow precise formal studies or even serve as efficient support for the realization of facsimiles.

Recognition of the technical process and the superpositions between figures. In rock art, studies are conditioned by the state of conservation of the figures, and the reading of superimpositions between strokes or figures is one of the fundamental problems. The system used has allowed the recognition of superpositions in at least three areas: (a) between lines of the same figure in cases where the composition of coloring matter is different; (b) between engraving and painting, as each technical action is spectrally discriminated; and, to a lesser extent, (c) superposition between engraved lines of the same figure, which allows reconstruction with a certain reliability of the process of execution of a motif.

Characterization of the state of conservation of cave motifs. The main utility provided by hyperspectral data is the possibility of carrying out exhaustive documentation of the conservation of a motif, perfectly discriminating between veiled areas, leached areas, scaling and any other action of a taphonomic nature to which the figure is subject. For the first time, it is now possible to present "maps" of the figures and the associated conservation problems. Once this long-term methodology has been applied and evaluated for the same reason, it will allow the study of the actions, frequency and intensity of each one, to obtain an exhaustive register that is above subjective and imprecise descriptions.

In conclusion, it can be noted that the application of hyperspectral remote sensing technology to the field of archaeology, and more specifically rock art, has high potential for applications in documentation, technical analysis and conservation of cultural heritage.

Author Contributions: Conceptualization, V.B. and S.R.; methodology, V.B.; software, V.B.; validation, S.R., M.A.S.; formal analysis, M.A.S.; funding acquisition V.B.; investigation, V.B., M.A.S. and S.R.; resources, V.B.; writing—original draft preparation, V.B.; writing—review and editing, V.B. and S.R.; visualization, V.B. and S.R.; supervision, M.A.S. and S.R.

Funding: This research was funded by Consejería de Innovación, Industria, Turismo y Comercio of Cantabrian Government in the context of aid to encourage industrial research and innovation in companies, project "Sistema para la Gestión y Difusión de Arte Rupestre y Parietal (SIGEDARP)," grant number IV07–XX-031.

Acknowledgments: The present paper has been produced within the scope of doctoral activities carried out by the lead author at the International Doctorate School of the Spanish National Distance-Learning University (Escuela Internacional de Doctorado de la Universidad Nacional de Educación a Distancia; EIDUNED). The authors are grateful for the support provided by this institution. The authors of this work would also like to thank the GIM Geomatics company for its support of the project and E. Castillo and R. Ontañón for their advice.

Conflicts of Interest: The authors declare no conflicts of interest.

References

1. Ontañon, R.; Bayarri, V.; Herrera, J.; Gutierrez, R. The conservation of prehistoric caves in Cantabria, Spain. In *The Conservation of Subterranean Cultural Heritage*; CRC Press/Balkema: Boca Raton, FL, USA; Taylor & Francis Group: London, UK, 2014; ISBN1 978-1-138-02694-0 (Hbk). ISBN2 978-1-315-73997-7.

2. Montero Ruiz, I.; Rodríguez Alcalde, Á.L.; Vicent García, J.M.; Cruz Berrocal, M. Técnicas digitales para la elaboración de calcos de arte rupestre. *Trabajos de Prehistoria* **2010**, *55*, 155–169. [CrossRef]

3. Ghamisi, P.; Yokoya, N.; Li, J.; Li, W.; Liu, S.; Plaza, J.; Rasti, B.; Plaza, A. Advances in Hyperspectral Image and Signal Processing: A Comprehensive Overview of the State of the Art. *IEEE Geosci. Remote Sens. Mag.* **2018**, *5*, 34–78. [CrossRef]

4. Hou, M.; Cao, N.; Tan, L.; Lyu, S.; Zhou, P.; Xu, C. Extraction of Hidden Information under Sootiness on Murals Based on Hyperspectral Image Enhancement Applied Sciences. *Appl. Sci.* **2019**, *9*, 3591. [CrossRef]

5. Cerra, D.; Agapiou, A.; Cavalli, R.M.; Sarris, A. An Objective Assessment of Hyperspectral Indicators for the Detection of Buried Archaeological Relics. *Remote Sens.* **2018**, *10*, 500. [CrossRef]

6. Goetz, A.F.H. Spectroscopic remote sensing for geological applications. *Imaging Spectrosc.* **1981**, *268*, 17–21.

7. Zhang, B. Advancement of hyperspectral image processing and information extraction. *J. Remote Sens.* **2016**, *20*, 1062–1090. [CrossRef]

8. Fischer, C.; Kakoulli, I. Multispectral and hyperspectral imaging technologies in conservation: Current research and potential applications. *Stud. Conserv.* **2013**, *51*, 3–16. [CrossRef]

9. Attas, M.; Cloutis, E.; Collins, C.; Goltz, D.; Majzels, C.; Mansfield, J.R.; Mantsch, H.H. Near-infrared spectroscopic imaging in art conservation: Investigation of drawing constituents. *J. Cult. Herit.* **2003**, *4*, 127–136. [CrossRef]

10. Bayarri, V.; Latova, J.; Castillo, E.; Lasheras, J.A.; De Las Heras, C.; Prada, A. Nueva documentación y estudio del arte empleando técnicas hiperespectrales en la Cueva de Altamira. ARKEOS | perspectivas em diálogo, n° 37. In *XIX International Rock Art Conference IFRAO 2015. Symbols in the Landscape: Rock Art and its Context. Conference Proceedings*; Instituto Terra e Memória: Tomar, Portugal, 2015; ISBN 978-84-9852-463-5.

11. Hauke, K.; Kehren, J.; Böhme, N.; Zimmer, S.; Geisler, T. In Situ Hyperspectral Raman Imaging: A New Method to Investigate Sintering Processes of Ceramic Material at High-temperature. *Appl. Sci.* **2019**, *9*, 1310. [CrossRef]

12. Alcalde Del Río, H.; Breuil, H.; Sierra, L. *Les Cavernes de la Région Cantabrique (Espagne)*; Impr. V A. Chêne: Cantabria, Monaco, 1911.

13. Ripoll Perelló, E. Nota acerca de algunas nuevas figuras rupestres de las cuevas de EL Castillo y La Pasiega (Puente Viesgo, Santander). In *Actas del IV Congreso Internacional de Ciencias Prehistóricas y Protohistóricas (Madrid 1954)*; Pan American Institute of Geography and History: Zaragoza, Spain, 1956; pp. 301–310.

14. García-Guinea, M.A.; González-Echegaray, J. Nouvelles reprèsentations d´art rupestre dans la grotte del Castillo. *Préhistoire et Spéléologie Ariégoises* **1966**, *XXI*, 441–446.

15. Echegaray, J.G.; Romanillo, J.A.M. Figuras rupestres inéditas de la Cueva del Castillo (Puente Viesgo). *Boletín del Seminario de Estudios de Arte y Arqueología* **1970**, *36*, 441–446.

16. Romanillo, J.A.M.; Sainz, C.G. Cronología del arte paleolítico cantábrico: Últimas aportaciones y estado actual de la cuestión. In *Actas del 3er Congreso de Arqueología Peninsular*; ADECAP: Vila Real, Portugal, 2000; Volume II, pp. 461–473.

17. Gárate-Maidagán, D. *Nuevos datos en torno al inicio del arte parietal cantábrico: La aportación de un caballo inédito en el Panel de Manos de la Cueva del Castillo (Puente Viesgo, Cantabria)*; Sautuola XII: Instituto de Prehistoria y Arqueología Sautuola: Santander, Spain, 2006; pp. 351–358.

18. Ruiz-Redondo, A. Una nueva revisión del Panel de las Manos de la cueva de El Castillo (Puente Viesgo, Cantabria). Munibe Antropologia-Arkeologia 61, C. Aranzadi. Z. E. Donostia/San Sebastián. 2010; 17–27.

19. Ripoll, S.; Bayarri, V.; Muñoz, F.J.; Latova, J.; Gutiérrez, R.; Pecci, H. El arte rupestre de la cueva de El Castillo (Puente Viesgo, Cantabria): Unas reflexiones metodológicas y una propuesta cronológica. In *Capítulo en Libro Cien Años de Arte Rupestre Paleolítico Centenario del Descubrimiento de la Cueva de la Peña de Candamo (1914–2014)*; Ediciones Universidad de Salamanca: Salamanca, Spain, 2014; ISBN 978-84-9012-480-2.

20. Kruse, F.A.; Lefkoff, A.B.; Boardman, J.W.; Heidebrecht, K.B.; Shapiro, A.T.; Barloon, P.J.; Goetz, A.F.H. The Spectral Image Processing System (SIPS)—Interactive visualization and analysis of imaging spectrometer data. *Remote Sens. Environ.* **1992**, *44*, 145–163. [CrossRef]

21. ICOM. *ICOM Guidelines for Loans*; International Council of Museums: Paris, France, 1994; Available online: https://icom.museum/wp-content/uploads/2018/07/Loans1974eng.pdf (accessed on 29 October 2019).

22. Green, A.A.; Berman, M.; Switzer, P.; Craig, M.D. A transformation for ordering multispectral data in terms of image quality with implications for noise removal. *IEEE Trans. Geosci. Remote Sens.* **1988**, *26*, 65–74. [CrossRef]

23. Rashmi, S.; Swapna, A.; Venkat Ravikiran, S. Spectral Angle Mapper Algorithm for Remote Sensing Image Classification. *Int. J. Innov. Sci. Eng. Technol.* **2014**, *1*, 2348–7968.

24. Sabins, F.F., Jr. *Remote Sensing Principles and Interpretation*, 2nd ed.; W. H. Freeman and Co: San Francisco, CA, USA, 1986; ISBN 13: 9780716717935.

25. Richards, J.A. *Remote Sensing Digital Image Analysis: An Introduction*; Springer: Berlin, Germany, 1999; p. 240.

26. Loève, M. *Probability Theory II*, 4th ed.; Graduate Texts in Mathematics. 46; Springer: New York, NY, USA, 1978; ISBN 978-0-387-90262-3.

27. Hyvärinen, A. Independent component analysis in the presence of gaussian noise by maximizing joint likelihood. *Neurocomputing* **1998**, *22*, 49–67. [CrossRef]

28. Hyvärinen, A. Fast and robust fixed-point algorithms for independent component analysis. *IEEE Trans. Neural Netw.* **1999**, *10*, 626–634. [CrossRef] [PubMed]

*applied sciences*

Article

Use of 3D Printing in Model Manufacturing for Minor Surgery Training of General Practitioners in Primary Care

M.C. Luque [1,2], A. Calleja-Hortelano [1] and P.E. Romero [1,*]

[1] Department of Mechanical Engineering, University of Cordoba, Medina Azahara Avenue, Cordoba 14071, Spain. luquerodriguezcarmen@gmail.com (M.C.L.); arcaho@gmail.com (A.C.-H.)

[2] Primary Care Center of La Carlota, La Paz Avenue, La Carlota, Cordoba 14100, Spain

* Correspondence: p62rocap@uco.es

Received: 1 November 2019; Accepted: 28 November 2019; Published: 30 November 2019

Featured Application: The proposed models can be used to train general practitioners in minor surgery courses in primary care centers.

Abstract: In order to increase the efficiency of the Spanish health system, minor surgery programs are currently carried out in primary care centers. This organizational change has led to the need to train many general practitioners (GPs) in this discipline on a practical level. Due to the cost of the existing minor surgery training models in the market, pig's feet or chicken thighs are used to practice the removal of figured lesions and the suture of wounds. In the present work, the use of 3D printing is proposed, to manufacture models that reproduce in a realistic way the most common lesions in minor surgery practice, and that allow doctors to be trained in an adequate way. Four models with the most common dermal lesions have been designed and manufactured, and then evaluated by a panel of experts. Face validity was demonstrated with four items on a five-point Likert scale that was completed anonymously. The models have obtained the following results: aesthetic recreation, 4.6 ± 0.5; realism during anesthesia infiltration, 4.8 ± 0.4; realism during lesion removal, 2.8 ± 0.4; realism during surgical wound closure, 1.2 ± 0.4. The score in this last section could be improved if a more elastic skin-colored filament were found on the market.

Keywords: 3D printing; additive manufacturing; fused deposition modeling (FDM); minor surgery; primary care; surgical training

---

1. Introduction

Additive manufacturing is a manufacturing process by which parts are generated directly from a three-dimensional (3D) model, usually using a single machine (3D printer). These 3D printers manufacture the part layer by layer, using a raw material (resin, powder, filament) and an energy source [1]. The fused deposition modeling (FDM) 3D printing technique is used in different sectors [2]: automotive, aeronautics, medicine, architecture, or art.

There are several 3D printing techniques [3]: stereolithography (SLA), polyjet (PJ), selective laser sintering (SLS), and binder jetting (BJ). However, the most commonly used technique today is the fused deposition modeling (FDM) technique [4–7]. This technique has several advantages [8]: machines and filaments have low costs; it is easy to find information on the internet about this technique; the learning curve is short; and there is a large catalog of filaments available on the market, with multiple qualities and applications [9].

This paper focuses on the use of 3D printing in primary care centers. For years, the FDM technique has been used at the hospital level in different services and areas, with different objectives [10,11]: (i) to train students and residents during their training period [12–14]; (ii) to explain the interventions to patients before the operation [15,16]; (iii) to prepare the intervention, when it is complex [17,18]; (iv) to manufacture custom prostheses [6], among other uses. Despite the increasing use of 3D printing in hospitals [19], the authors have not found any work related to the use of 3D printing in primary care. In the present work, we propose the use of 3D printing to train general practitioners in minor surgery.

In Spain, with the development of primary care, the competencies of general practitioners (GPs) have been expanded with the aim of increasing the efficiency of the health system [20]. As in other European countries, the development of one of these competencies led to the creation of minor surgery programs in health centers [21]. In them, interventions for common lesions (dermal nevus, seborrheic keratosis, or epidermal cyst) are carried out [22].

GPs should learn these skills during their resident internal doctor-training period. However, due to the shortage of skilled adjunct professionals, practical training in this field is not adequate [23]. To make up for this lack, GPs interested in developing these aspects of their professional competence have to be trained through specific courses [24].

In minor surgery courses, the different phases of an intervention are taught [25]: (i) infiltration of anesthesia at the local level, (ii) removal of the lesion, and (iii) closure of the surgical wound. There are training models on the market that allow some of these phases to be practiced; however, these models have several handicaps: they do not have all the typical lesions in minor surgery, they do not faithfully reproduce the lesions of minor surgery, and they have a high cost. For these reasons, courses traditionally use pig's feet, chicken thighs or bacon to practice the removal of figured lesions and suturing of wounds.

The purpose of this project is to design, manufacture (via the FDM 3D printing technique), and evaluate models for minor surgery courses in primary care centers. For this purpose, the most common lesions in minor surgery have been selected and modeled in 3D. These models have been manufactured using a double extruder printer, equipped with flexible filament. The models thus manufactured have been evaluated by a panel of experts in minor surgery. In addition, the models have been uploaded to an open standard tessellation (STL) file-exchange platform so that they can be used or improved by any interested person.

2. Materials and Methods

The methodology followed during this work has been the following (Figure 1): (i) in a first phase, it was decided which lesions were going to be modeled, the photographs to be taken as reference were selected, and some preliminary sketches were elaborated; (ii) then, the different lesions were modeled, using a parametric software of 3D design; (iii) the third phase consisted of importing the STL file from the slicing software, and selecting the print parameters necessary to generate the corresponding numerical control program; (iv) the different lesions were then printed; (v) once manufactured, the models were evaluated by a panel of experts; (vi) finally, the models were uploaded to an STL file-exchange platform.

2.1. Initial Stage

The first stage of the work corresponds to a series of meetings of the research team. The first decision made during this stage was which lesions were to be modeled (Figure 2): seborrheic keratosis, epidermal cyst, dermal nevus, and ingrown toenail. These lesions were chosen because they are the most frequently occurring in minor surgery patients.

During this stage, sketches of the lesions were drawn. In addition, important decisions were reached about the size of the models, the most suitable position for printing, the entities that make up the model, and use of the different materials or densities needed. These decisions are important as they have a direct impact on the manufacturing process (Table 1) [26]. At this stage, it was also agreed to fix the models to the table, so that they are firmly tied to the table during the intervention simulations.

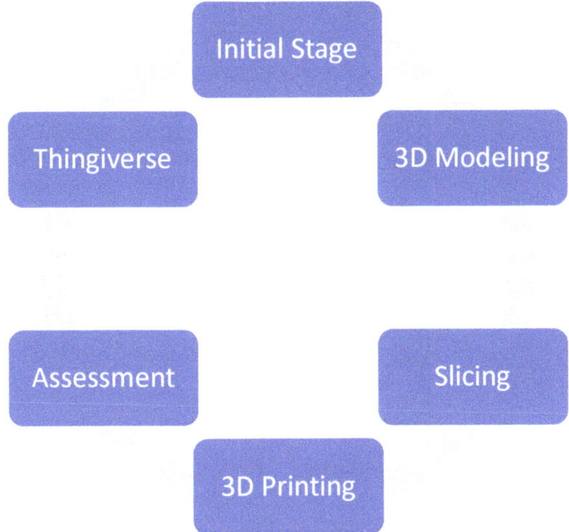

Figure 1. Diagram summarizing graphically the stages followed during the execution of the work: initial phase, where the sketches of the models were generated; the 3D modeling of the different models; slicing of the models and configuration of the 3D printer; the 3D printing of the models and post-processing tasks; evaluation of the models; uploading of the standard tessellation (STL) files to Thingiverse.

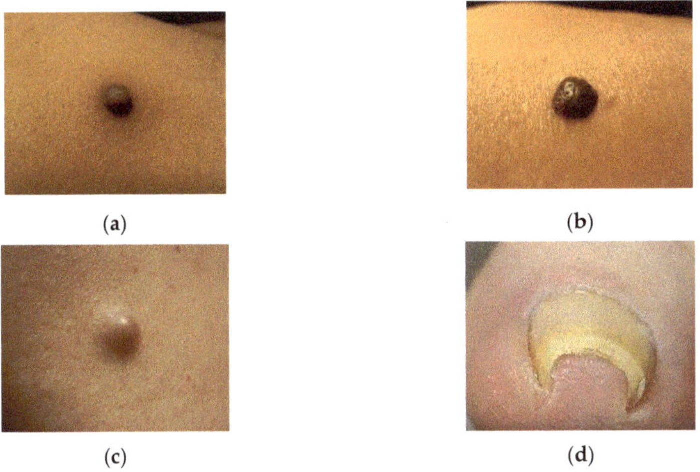

(a)

(b)

(c)

(d)

Figure 2. Images of the most common lesions in the minor surgery practice: dermal nevus (**a**); seborrheic keratosis (**b**); epidermal cyst (**c**); ingrown toenail (**d**).

Table 1. Relationship between initial decisions and the designing and manufacturing process.

Initial Decisions	Have Influence on
Size	Print time
Print position	Possibility of printing Surface roughness
Entities that make up the model	Procedure to model the lesion in 3D
Combination of materials	Using the two 3D printer extruders
Print density	Flexibility of the model Print time
Pattern	Flexibility of the model

2.2. 3D Modeling

From photographs (Figure 2), and from the sketches created in the previous stage (Figure 3a), 3D models of the lesions were made. The parametric software SolidWorks [27] was used for this purpose (Figure 3b). Once the lesion was modeled, the file was exported in STL standard, which is the format accepted by the slicing software.

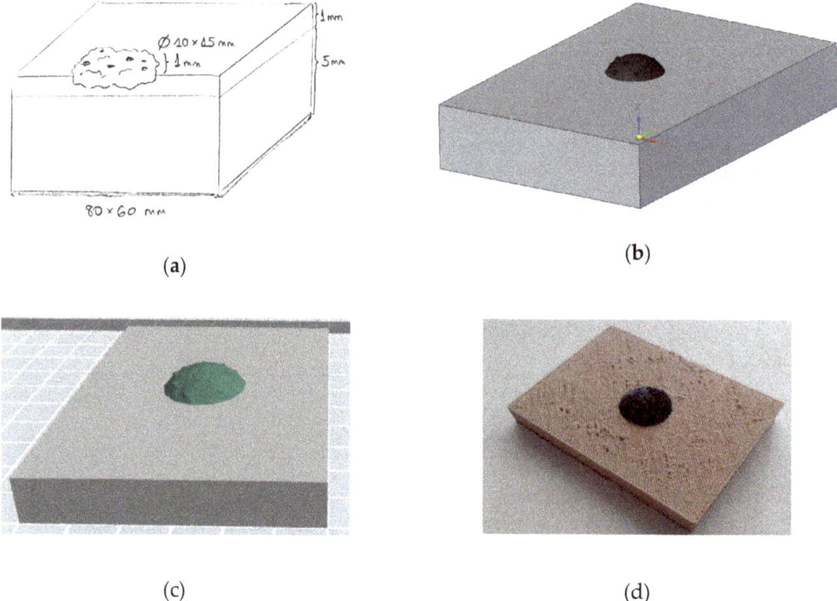

(a)

(b)

(c)

(d)

Figure 3. Modeling and manufacturing of the seborrheic keratosis training model: draft developed in initial stage (**a**); the 3D model generated by SolidWorks software (**b**); the imported model in FlashPrint slicing software (**c**); the 3D printed lesion (**d**).

The models corresponding to the dermal nevus, seborrheic keratosis and epidermal cyst were modeled with dimensions of $80 \times 60 \times 6$ mm^3. The model to reproduce the ingrown toenail was modeled to a realistic size.

2.3. Slicing

A numerical control (NC) code, which tells 3D printers what to do at any given moment, was needed. These NC codes are generated by slicing software from the STL files (Figure 3c). The 3D printer used in this work has been manufactured by FlashForge [28]. This company has developed its own slicing software for its 3D printers called FlashPrint [29]. In this software, it is necessary to indicate which material is going to be used, and which values for printing parameters are going to be selected [8]: extrusion temperature, layer height, print speed, among others. The values used in the present work are shown in Table 2.

Table 2. Parameter values used during printing.

Print Parameter	Value
Extrusion Temperature	225 °C
Bed Temperature	50 °C
Layer Height	0.18 mm
First-Level Layer Height	0.27 mm
Retraction Length	3.7 mm
Fill Density	90%
Fill Pattern	Hexagon

2.4. 3D Printing

A FlashForge Creator Pro printer was used for the printing process. This machine has a hot bed (dimensions 225 x 125 mm^2) and is equipped with two extruders (Figure 4). Several filaments have been used to print the models: a skin-resembling thermoplastic elastomer (TPE), Recreus brand (for the skin); a skin-resembling TPE, Tianse brand (for the inner sphere of the epidermal cyst and the nail); and a black TPE, Smart Materials 3D (for the seborrheic keratosis, Figure 3d).

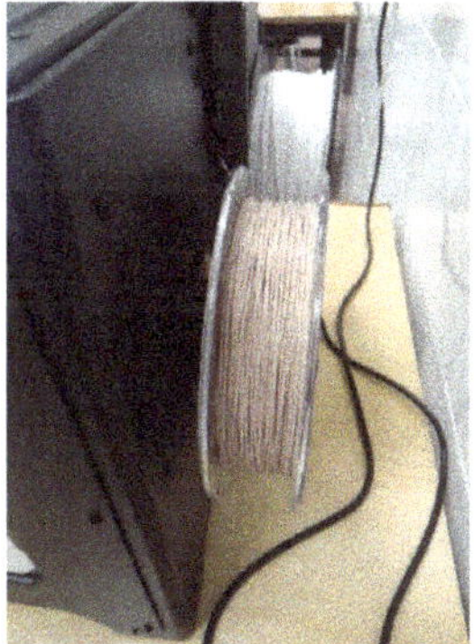

Figure 4. FlashForge Creator Pro 3D printer used in the project (**left**); rear of the printer, with two bobbins of skin-resembling thermoplastic elastomer (TPE) filament (**right**).

2.5. Assessment

Once the models were printed, they underwent evaluation by a panel made up of five experts in minor surgery. The members of the evaluation panel are general practitioners, who work in rural primary care centers belonging to the Andalusian Health Service, and who each have more than 15 years of experience. These general practitioners perform interventions at least once a week.

It is difficult to find general practitioners who are experts in minor surgery. Although the number of experts may seem insufficient, according to [30] the results obtained with the use of a panel of five experts are adequate.

Each expert performed an operation simulation using each of the models. Subsequently, the experts filled out a survey, wherein they were asked their opinion on four main aspects of realism in the training models: aesthetics, infiltration of the anesthesia, removal of the lesion, and closure of the surgical wound. In this survey, the Likert 5-point scale was used (1—very disagreeable; 5—very agreeable). For additional information, the experts wrote a paragraph in relation to each of the aspects evaluated.

2.6. Sharing the Models

After the evaluation, the STL models were uploaded to Thingiverse [31]. Thingiverse is a repository for sharing 3D pieces. The idea is to share the designs developed during this project through creative commons license, so that anyone interested in performing practice in minor surgery can print them at home or at a primary care center.

3. Results

In the present work, models that reproduce lesions typical of minor surgery were designed and manufactured using 3D printing in order to train GPs in practical courses in this discipline. Once printed, the models were evaluated by a panel of experts. The aspects evaluated were: the aesthetic recreation of the models, realism during the infiltration of the anesthesia, realism in the removal of the lesion, and realism during the closure of the surgical wound. After the evaluation, the models were uploaded to Thingiverse [31] so that they can be used by any interested user (Figure 5).

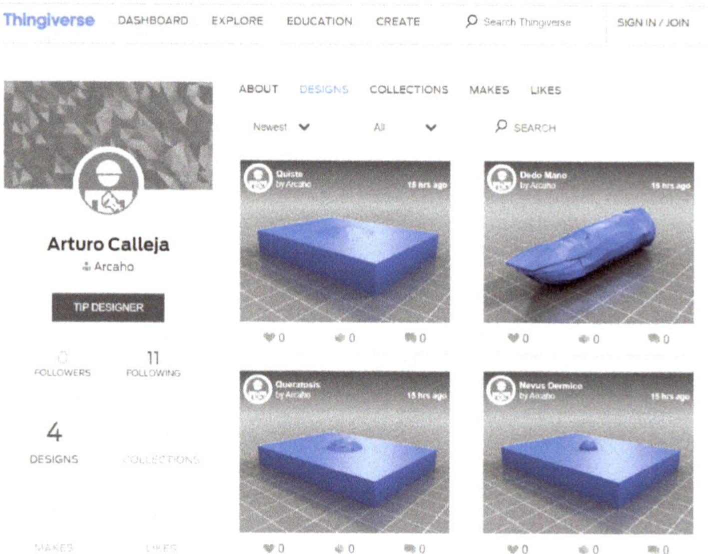

Figure 5. User web page created in Thingiverse to upload the models of minor surgery injuries. Any interested user can download these standard tessellation (STL) models for free and print them in their home or primary care center.

3.1. Aesthetic Recreation Level

The score obtained by the models in this aspect was 4.6 ± 0.5. The panel of experts agreed that the models realistically recreate the most common injuries in minor surgery (Figure 6). One of the experts especially valued the effort made, as he said that the work is not easy. Another expert confessed that this work opens a very hopeful door to the teaching of minor surgery, which is as important as it is unknown to most GPs working in the field of primary care.

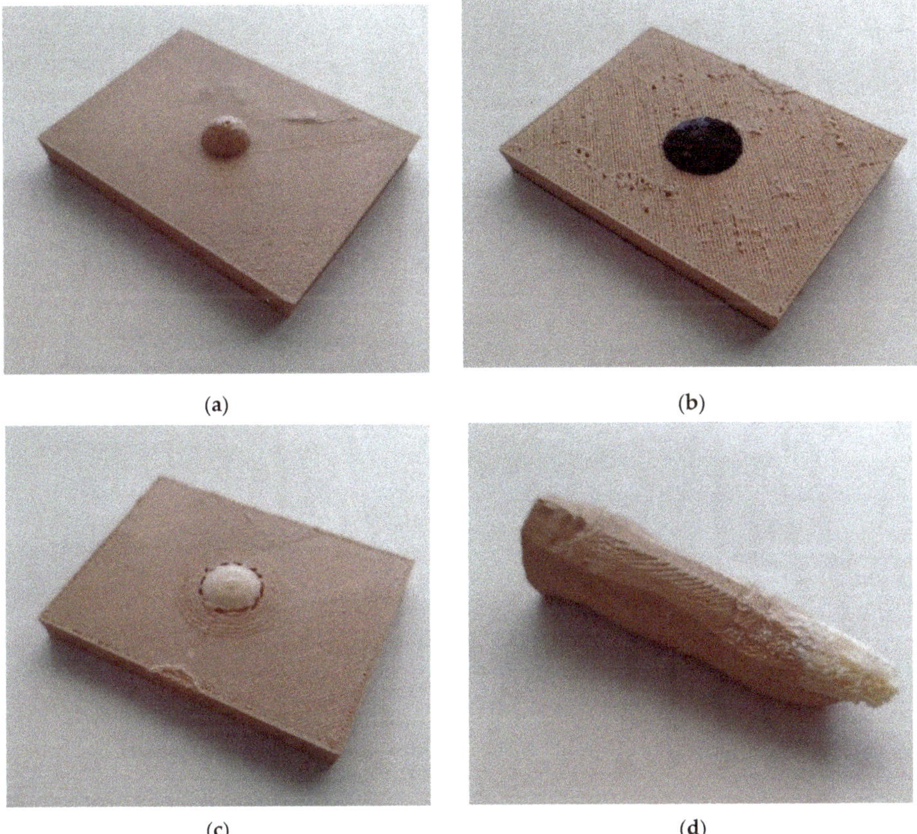

Figure 6. Models of minor surgical lesions: dermal nevus (**a**); seborrheic keratosis (**b**); epidermal cyst (**c**); ingrown toenail (**d**).

3.2. Realism During Anesthesia Infiltration

The score obtained by the models in this aspect was 4.8 ± 0.4. On this issue, the panel of experts agreed that the feel of the infiltration realistic (Figure 7a,b); it is especially realistic in the lesion of seborrheic keratosis and in the toe. One of the experts indicated that an improvement of the models would involve the possibility of injecting water with red dye to simulate blood, if not in the entire model, then at least around the lesion. In the case of the toe, a rigid tube could be added in the center to simulate the bone and a small 'pipe' on both sides to simulate the blood vessels.

3.3. Realism During Removal of the Lesion

The score obtained by the models in this aspect was 2.8 ± 0.4. In this third question, the panel of experts agreed again: in general, the sense of touch is not very similar to reality (Figure 7c). The material used is harder than human skin. There is even the risk of cutting yourself with the scalpel. One of the experts recommended using a material like silicone.

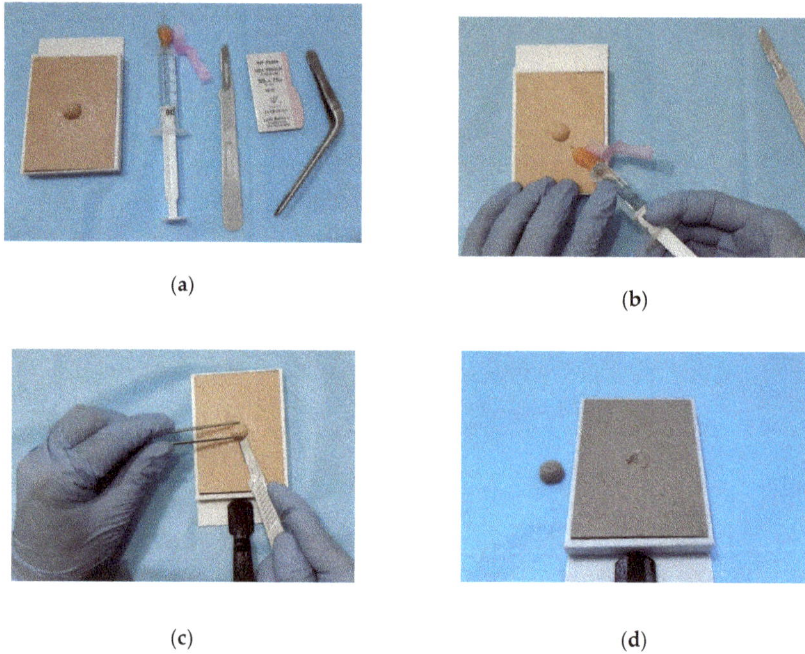

Figure 7. Stages of the intervention performed on the dermal nevus model: preparation of the necessary material (**a**); infiltration of the anesthesia (**b**); removal of the lesion (**c**); outcome (**d**).

On the other hand, in the lesion of the cyst, the 'skin' should completely cover the cyst capsule. It is necessary to paint it to delimit the lesion and it is also necessary to have superficial material so that once the capsule has been removed, the skin can be sutured (bringing the edges of the wound closer together).

3.4. Realism During Surgical Wound Closure

The score obtained by the models in this aspect was 1.2 ± 0.4. In the case of seborrheic keratosis and fibroepithelioma, the wound would not be sutured but cauterized with the electric scalpel (Figure 7d). In the case of the cyst and toe, the silk suture would be used. However, the material is too hard, and the suture is very difficult to perform. The toe and nail do not reproduce the lesion well, so their intervention is not viable. The nail should be a thinner sheet and in the proximal part should be included 'inside the skin'.

3.5. Cost of Models

The technology of fused filament deposition modeling has a low cost: printers are becoming cheaper and have a low electrical consumption; on the other hand, filaments cost between 25 and 60 euros/kilo, depending on the type of material.

From the data shown in Table 3, the cost of the different models manufactured has been calculated (Table 4). The price of labor has not been included in this calculation, as it is understood that the supervision of a printer could be carried out by resident internal doctors as part of their training. The price of each model is around 6 Euros (the ingrown toenail is cheaper because it has less material). The cost of a commercial training model is 38 Euros.

Table 3. Parameter values used during costing.

Parameter	Value
Cost of plastic (€/kg)	60
Cost of electrical energy (€/kWh)	0.15
3D Printer power (kW)	0.500
Cost of the 3D printer (€)	599
Amortization period (years)	1
Days active per year	250
Hours per day (h)	8
Failure rate (%)	5

Table 4. Cost of printing the training models manufactured (not including the cost of labor, as the printing could be supervised by internal doctors' residents as part of their training process).

Parameter	Dermal Nevus	Seborrheic Keratosis	Epidermal Cyst	Ingrown Nail
Workpiece mass (kg)	0.06	0.05	0.06	0.01
Printing time (h)	5.9	5.80	6.00	2.01
Cost of plastic material (€)	3.61	3.10	3.61	0.98
Cost of electricity (€)	0.44	0.44	0.45	0.15
Equipment amortization cost (€)	1.77	1.74	1.80	0.60
Cost of failures (€)	0.29	0.26	0.29	0.09
Total (€)	6.11	5.53	6.15	1.81

4. Discussion

In this work, 3D models that reproduce the most frequent lesions in minor surgery have been modeled, manufactured and evaluated. These models are representative of realistic lesions, since they are intended to be used to train GPs in primary care centers.

Manufactured training models cost less than commercial ones and faithfully reproduce injuries and lesions. In addition, they have been developed under an open-source license, so that anyone interested can download and print them from their home or primary care center.

The models have been evaluated by a panel of experts in minor surgery. Four aspects were assessed: the level of aesthetic recreation, realism during infiltration of the anesthesia, realism during removal of the lesion, and realism during closure of the surgical wound.

The panel of experts valued the level of recreation of the lesions positively, as well as the realism during the infiltration of the anesthesia. The removal of the lesion is more laborious than in real interventions, due to the density selected as the printing parameter and the qualities of the material used. It is not possible to close the surgical wound because the material is not elastic enough.

From the results obtained, the following statements can be made:

- Primary care centers can now access 3D printing. In hospitals, 3D-printed models have been used for several years to train surgeons [11,19]. Waran et al. [17] have found that the use of training models manufactured using 3D printing increases the success rate in major surgery operations and reduces the time spent on each intervention. Driven by this reality, the authors propose the use of this methodology in primary care centers, to carry out practical courses that serve to initiate GPs in minor surgery.
- The cost of the proposed models is low. FDM 3D printing allows healthcare resources to be manufactured at a low cost [32]. The training models proposed in this work have a cost of 6 Euros, compared to 38 Euros for commercial training models. This detail is important in a public health system such as the Spanish one, or in developing countries [33]. In addition, the proposed process allows a user to customize the models, and to manufacture those that have interest in each geographic point [34].
- The proposed models have a high level of aesthetic recreation. In major surgery, when lesions have some complexity, images taken by computed tomography (CT) and magnetic resonance imaging

(MRI) are used to model the lesion in 3D [35]. In this case, the models have been performed in parametric design software from photographs and drawings made by one of the authors, an expert in minor surgery. However, the panel of experts have positively assessed the level of aesthetic recreation in the models.

- The models have been shared on Thingiverse. In the 'maker' world it is common to share designed models. This is one of the main attractions of 3D printing. Thingiverse is a widely used platform for sharing general models, although there are specific platforms for biomedical models, such as one developed by the National Institutes of Health of United States (NIH 3D) [36].

- FDM 3D printing technology is suitable for this purpose. There are 3D printing technologies that can print models of great complexity [37], although their cost cannot be borne by a primary care center. In the present work, the use of FDM technology has been proposed, which is the cheapest technology, and which provides a surface finish suitable for the intended use.

- The filament used is not elastic enough and it is not possible to close the surgical wound. It is necessary to look in the market for more elastic filaments that are available in skin color to overcome this handicap.

5. Conclusions

The present work proposes the use of the FDM technique of 3D printing to manufacture models of typical lesions in minor surgery. It is intended to use these models to train general practitioners in primary care centers. The models have been manufactured and evaluated by a panel of experts. The models have been approved in three of the four categories evaluated. In addition, the models have been uploaded to Thingiverse for anyone to download and use.

Author Contributions: Conceptualization, M.C.L. and P.E.R.; methodology, M.C.L. and A.C.-H.; software, A.C.-H.; validation, M.C.L., A.C.-H. and P.E.R.; formal analysis, M.C.L.; investigation, M.C.L.; resources, P.E.R.; data curation, A.C.-H.; writing—original draft preparation, M.C.L.; writing—review and editing, P.E.R.; visualization, M.C.L. and A.C.-H.; supervision, P.E.R.; project administration, P.E.R.; funding acquisition, P.E.R.

Funding: The authors would like to thank the funding received by the University of Cordoba's Own Research Plan (2019).

Conflicts of Interest: The authors declare no conflicts of interest.

References

1. Lipson, H.; Kurman, M. *Fabricated: The New World of 3D Printing*; John Wiley & Sons Inc.: Indianapolis, IN, USA, 2013.
2. Wong, K.V.; Hernandez, A. A Review of Additive Manufacturing. *ISRN Mech. Eng.* **2012**, *2012*, 1–10. [CrossRef]
3. Sun, Z.; Lau, I.; Wong, Y.H.; Yeong, C.H. Personalized Three-Dimensional Printed Models in Congenital Heart Disease. *J. Clin. Med.* **2019**, *8*, 522. [CrossRef] [PubMed]
4. Ahn, S.; Montero, M.; Odel, D.; Roundy, S.; Wright, P.K. Anisotropic material properties of fused deposition modeling ABS. *Rapid Prototyp. J.* **2002**, *8*, 248–257. [CrossRef]
5. Ngo, T.D.; Kashani, A.; Imbalzano, G.; Nguyen, K.T.Q.; Hui, D. Additive manufacturing (3D printing): A review of materials, methods, applications and challenges. *Compos. Part B Eng.* **2018**, *143*, 172–196. [CrossRef]
6. Ten Kate, J.; Smit, G.; Breedveld, P. 3D-printed upper limb prostheses: A review. *Disabil. Rehabil. Assist. Technol.* **2017**, *12*, 300–314. [CrossRef]
7. Turner, B.N.; Strong, R.; Gold, S.A. A review of melt extrusion additive manufacturing processes: I. Process design and modeling. *Rapid Prototyp. J.* **2014**, *20*, 192–204. [CrossRef]
8. Barrios, J.M.; Romero, P.E. Improvement of surface roughness and hydrophobicity in PETG parts manufactured via fused deposition modeling (FDM): An application in 3D printed self—Cleaning parts. *Materials* **2019**, *12*, 2499. [CrossRef]
9. Smart Materials 3D. Available online: https://www.smartmaterials3d.com/en/ (accessed on 17 January 2019).

10. Lee Ventola, C. Medical applications for 3D printing: Current and projected uses. *Pharm. Ther.* **2014**, *39*, 704–711.

11. Tack, P.; Victor, J.; Gemmel, P.; Annemans, L. 3D-printing techniques in a medical setting: A systematic literature review. *Biomed. Eng. Online* **2016**, *15*, 115. [CrossRef]

12. Mcmenamin, P.G.; Quayle, M.R.; Mchenry, C.R.; Adams, J.W. The production of anatomical teaching resources using three-dimensional (3D) printing technology. *Anat. Sci. Educ.* **2014**, *7*, 479–486. [CrossRef]

13. Barber, S.R.; Kozin, E.D.; Dedmon, M.; Lin, B.M.; Lee, K.; Sinha, S.; Black, N.; Remenschneider, A.K.; Lee, D.J. 3D-printed pediatric endoscopic ear surgery simulator for surgical training. *Int. J. Pediatr. Otorhinolaryngol.* **2016**, *90*, 113–118. [CrossRef] [PubMed]

14. Ho, D.; Squelch, A.; Sun, Z. Modelling of aortic aneurysm and aortic dissection through 3D printing. *J. Med. Radiat. Sci.* **2017**, *64*, 10–17. [CrossRef] [PubMed]

15. Pérez-Mañanes, R.; Calvo-Haro, J.; Arnal-Burró, J.; Chana-Rodríguez, F.; Sanz-Ruiz, P.; Vaquero-Martín, J. Our experience with domestic 3D printing in Orthopedic Surgery and Traumatology. Do it yourself. *Rev. Latinoam. Cirugía Ortopédica* **2016**, *1*, 47–53. [CrossRef]

16. Allan, A.; Kealley, C.; Squelch, A.; Wong, Y.H.; Yeong, C.H.; Sun, Z. Patient-specific 3D printed model of biliary ducts with congenital cyst. *Quant. Imaging Med. Surg.* **2019**, *9*, 86–93. [CrossRef]

17. Waran, V.; Narayanan, V.; Karuppiah, R.; Pancharatnam, D.; Chandran, H.; Raman, R.; Rahman, Z.A.A.; Owen, S.L.F.; Aziz, T.Z. Injecting realism in surgical training—Initial simulation experience with custom 3D models. *J. Surg. Educ.* **2014**, *71*, 193–197. [CrossRef]

18. Michalski, M.H.; Ross, J.S. The shape of things to come: 3D printing in medicine. *JAMA* **2014**, *312*, 2213–2214. [CrossRef]

19. Martelli, N.; Serrano, C.; Van Den Brink, H.; Pineau, J.; Prognon, P.; Borget, I.; El Batti, S. Advantages and disadvantages of 3-dimensional printing in surgery: A systematic review. *Surgery* **2016**, *159*, 1485–1500. [CrossRef]

20. Tárraga López, P.J.; Marín Nieto, E.; García Olmo, D.; Celada Rodríguez, A.; Solera Albero, J. Economic impact of the introduction of a minor surgery program in primary care. *Aten. Primaria* **2001**, *27*, 335–338. [CrossRef]

21. Gokani, V.J.; Ferguson, H.J.M.; Fitzgerald, J.E.F.; Beamish, A.J. Surgical training in primary care: Consensus recommendations by the Association of Surgeons in Training. *Int. J. Surg.* **2014**, *12*, S1–S4. [CrossRef]

22. Serra, M.; Arevalo, A.; Ortega, C.; Ripoll, A.; Giménez, N. Minor surgery activity in primary care. *J. R. Soc. Med. Short Rep.* **2010**, *1*, 1–8. [CrossRef]

23. Puche, J.M.; Muñoz, P.A. Program to introduce and develop minor surgery at a primary-care center in Spain. *Arch. Med. Fam.* **2009**, *11*, 39–49.

24. Garrido, S.E.; García, R.V.; Nogales, P.A. Continuing education in primary care: The educational needs of its professionals. *Aten. Primaria* **2002**, *30*, 368–373.

25. Brown, J. *Minor Surgery, a Text and Atlas*, 4th ed.; CRC Press: London, UK, 2001; ISBN 978-0340761137.

26. Thompson, M.K.; Moroni, G.; Vaneker, T.; Fadel, G.; Campbell, R.I.; Gibson, I.; Bernard, A.; Schulz, J.; Graf, P.; Ahuja, B.; et al. Design for additive manufacturing: Trends, opportunities, considerations, and constraints. *CIRP Ann. Manuf. Technol.* **2016**, *65*, 737–760. [CrossRef]

27. SolidWorks. Available online: https://www.solidworks.com/ (accessed on 1 November 2019).

28. FlashForge. Available online: https://www.flashforge.com/ (accessed on 1 November 2019).

29. FlashPrint. Available online: https://www.flashforge.com/download-center (accessed on 1 November 2019).

30. Hyrkäs, K.; Appelqvist-Schmidlechner, K.; Oksa, L. Validating an instrument for clinical supervision using an expert panel. *Int. J. Nurs. Stud.* **2003**, *40*, 619–625. [CrossRef]

31. Thingiverse. Available online: https://www.thingiverse.com/ (accessed on 1 November 2019).

32. Rismani, S.; Van der Loos, H.F.M. The Competitive Advantage of Using 3D-Printing in Low-Resource Healthcare Settings. In Proceedings of the 20th International Conference on Engineering Design, Milan, Italy, 27–30 July 2015; pp. 1 10.

33. Corsini, L.; Aranda-Jan, C.B.; Moultrie, J. Using digital fabrication tools to provide humanitarian and development aid in low-resource settings. *Technol. Soc.* **2019**, *58*, 101117. [CrossRef]

34. Petrick, I.J.; Simpson, T.W. 3D printing disrupts manufacturing. *Res. Technol. Manag.* **2013**, *56*, 12–16. [CrossRef]

35. Rengier, F.; Mehndiratta, A.; Von Tengg-Kobligk, H.; Zechmann, C.M.; Unterhinninghofen, R.; Kauczor, H.U.; Giesel, F.L. 3D printing based on imaging data: Review of medical applications. *Int. J. Comput. Assist. Radiol. Surg.* **2010**, *5*, 335–341. [CrossRef]

36. Coakley, M.F.; Hurt, D.E.; Weber, N.; Mtingwa, M.; Fincher, E.C.; Alekseyev, V.; Chen, D.T.; Yun, A.; Gizaw, M.; Swan, J.; et al. The NIH 3D print exchange: A public resource for bioscientific and biomedical 3D prints. *3D Print. Addit. Manuf.* **2014**, *1*, 137–140. [CrossRef]

37. Jacobs, S.; Grunert, R.; Mohr, F.W.; Falk, V. 3D-Imaging of cardiac structures using 3D heart models for planning in heart surgery: A preliminary study. *Interact. Cardiovasc. Thorac. Surg.* **2008**, *7*, 6–9. [CrossRef]

Article

A Hybrid Machine Learning and Population Knowledge Mining Method to Minimize Makespan and Total Tardiness of Multi-Variety Products

Yongtao Qiu [1], Weixi Ji [1,2,*] and Chaoyang Zhang [1]

[1] School of Mechanical Engineering, Jiangnan University, Wuxi 214122, China; qyt_jiangnan@163.com (Y.Q.); zcy_jiangnan@outlook.com (C.Z.)

[2] Key Laboratory of Advanced Manufacturing Equipment Technology, Jiangnan University, Wuxi 214122, China

* Correspondence: jiweixi_jiangnan@outlook.com; Tel.: +86-139-2150-1567

Received: 6 November 2019; Accepted: 2 December 2019; Published: 4 December 2019

Abstract: Nowadays, the production model of many enterprises is multi-variety customized production, and the makespan and total tardiness are the main metrics for enterprises to make production plans. This requires us to develop a more effective production plan promptly with limited resources. Previous research focuses on dispatching rules and algorithms, but the application of the knowledge mining method for multi-variety products is limited. In this paper, a hybrid machine learning and population knowledge mining method to minimize makespan and total tardiness for multi-variety products is proposed. First, through offline machine learning and data mining, attributes of operations are selected to mine the initial population knowledge. Second, an addition–deletion sorting method (ADSM) is proposed to reprioritize operations and then form the rule-based initial population. Finally, the nondominated sorting genetic algorithm II (NSGA-II) hybrid with simulated annealing is used to obtain the Pareto solutions. To evaluate the effectiveness of the proposed method, three other types of initial populations were considered under different iterations and population sizes. The experimental results demonstrate that the new approach has a good performance in solving the multi-variety production planning problems, whether it is the function value or the performance metric of the acquired Pareto solutions.

Keywords: initial population; data mining; multi-variety; machine learning; production planning

---

1. Introduction

In recent years, digital integration and artificial intelligence have accelerated at an explosive rate. This progress inevitably leads to the rapid development of two technologies: big data [1–3] and intelligent algorithms [4]. For enterprise production planning or scheduling, although there are many machine learning algorithms [5–9] for this problem, there are still not many data mining methods, especially for multi-objective job shop scheduling problems (MOJSSP) [10].

MOJSSP is an important and critical problem for today's enterprises. A MOJSSP model can typically be described as a set of machines and jobs under multiple objectives, where each job has different sequential operations and each job should be processed on machines in a given order. This pattern is in line with the production planning problem for multi-variety products with different routings. Herein, the problem is transformed into how to rationally arrange the order of operations of multi-variety products on various machine tools, such that two or more objectives are optimal, e.g., makespan and total tardiness.

The characteristics of multi-objective problems increase the computational difficulty of traditional single-objective scheduling, while data mining [11] could extract rules from a large dataset without too much professional scheduling knowledge, which has more research space in solving MOJSSP and obtaining the optimal values of the objectives.

For the benefits of data mining, a multi-objective production planning method combining machine learning and population knowledge mining is proposed. Through the relation hierarchies mapping [12], attributes related to processing time and due date are selected. The machine learning algorithm used here is a hybrid metaheuristic algorithm that combines nondominated sorting genetic algorithm II (NSGA-II) and simulated annealing. It provides training data for the mining of population knowledge, and the rules obtained can be used as the criterion for sorting operations. Each operation gets its priority through rules, but it sometimes fails to meet the requirements of the rules (the number of operations per priority is fixed). Besides, the sequence of operations should also be considered. Therefore, this paper proposes a comprehensive priority assignment procedure called addition–deletion sorting method (ADSM) which can meet not only the optimized priority but also the requirements of rules and the order of operations. The main contributions of this paper are highlighted as follows:

(1)   A hybrid machine learning and population knowledge mining method is proposed to solve multi-objective job shop scheduling problems.
(2)   Five attributes, namely operation feature, processing time, remaining time, due date, and priorities, are selected to mine initial population knowledge.
(3)   The ADSM method is designed to reprioritize operations after the population knowledge mining.
(4)   Three populations (rules, mixed, and random) with different iterations and population sizes are compared, and three performance metrics are defined to explore the effectiveness of the proposed method.

2. Literature Review

MOJSSP is increasingly attracting scholars and researchers. Most methods are based on a multi-objective evolutionary algorithm (MOEA) [13], which first generates an initial population of production planning and then iterates continuously to obtain the best results. Huang [14] proposed a hybrid genetic algorithm to solve the scheduling problem while considering transportation time. Souier et al. [15] used NSGA-II for real-time scheduling under uncertainty and reliability constraints. Ahmadi et al. [16] used NSGA-II and non-dominated ranking genetic algorithm (NRGA) to optimize the makespan and stability under the disturbance of machine breakdown. Zhou et al. [17] presented a cooperative coevolution genetic programming with two sub-populations to generate scheduling policies. Zhang et al. [18] utilized a genetic algorithm combined with enhanced local search to solve the energy-efficiency job shop scheduling problems. A high-dimensional multi-objective optimization method was proposed by Du et al. [19]. To improve the robustness of scheduling, constrained nondominated sorting differential evolution based on decision variable classification is developed to acquire the ideal set.

In addition to the MOEA method, there are many other approaches to solve MOJSSP. Sheikhalishahi et al. [20] studied the multi-objective particle swarm optimization method. In their paper, makespan, human error, and machine availability are considered. Lu et al. [21] employed a cellular grey wolf optimizer (GWO) for the multi-objective scheduling problem with the objectives of noise and energy, while Qin et al. [22] tried to overcome the premature convergence of the GWO in solving MOJSSP by combining an improved tabu search algorithm. Zhou et al. [23] proposed three agent-based hyper-heuristics to solve scheduling policies with constrains of dynamic events. Fu et al. [24] designed a multi-objective brain storm optimization algorithm to minimize the total tardiness and energy consumption. Their goal is to minimize multiple objectives through MOEA or other metaheuristic algorithms, without exploiting a data mining method to improve the performance of scheduling.

With the development of industrial intelligence, data mining could be applied to management analysis and knowledge extraction. For MOJSSP, it could play a key role to solve the scheduling problem in the future. Recently, some researchers are trying to apply data mining to overcome the difficulties of job shop scheduling problems. Ingimundardottir and Runarsson [25] used imitation learning to discover dispatching rules. Li and Olafsson [26] introduced a method for generating scheduling rules using the decision tree. In their research, data mining extracts the rules that the attributes of jobs determine which one is better to be processed first. Similarly, Olafsson and Li [27] applied a decision tree to learn directly from scheduling data where the selected instances are identified as preferred training data. Jun et al. [28] proposed a random-forest-based approach to extract dispatching rules from the best schedules. This approach includes schedule generation, rule learning, and discretization such that they could minimize the average total weighted tardiness for job shop scheduling. Kumar and Rao [29] applied data mining to extract patterns in data generated by the ant colony algorithm, which generate a rule set scheduler that approximates the ant colony algorithm's scheduler. Nasiri et al. [30] gained a rule-based initial population via GES/TS method, and then employed PSO and GA to verify the advantage of this method, but lacked the consideration of operations sequence. Most of them considered data mining as a tool to extract dispatching rules, while only a few of them utilized a data mining approach to generate the initial population of metaheuristic methods, not to mention multi-objective job shop scheduling problem.

This paper uses data mining to generate the rule-based initial population for MOJSSP. The operation sequence is considered and an addition–deletion sorting method is presented to reprioritize the order of operations of each job. The advantage of the proposed method is reflected by the NSGA-II with simulated annealing method of the random initial population under different criteria of iteration numbers and different initial population sizes.

3. Problem Description and Goal

The MOJSSP can be described as n jobs $\{J_0, J_1, \ldots, J_n\}$ under multiple objectives processed on m machines $\{M_0, M_1, \ldots, M_m\}$ with different routes [31]. The processing order of each part and the processing time of each operation o_{ij} on a machine is determined. To illustrate the problem, the encoding of a benchmark of 10×10 job shop (LA18) [32] here can take an example as $\{9, 8, 5, 4, 2, 1, 3, 6, 7, 0, \ldots, 5, 8, 9\}$. Each element in the operation sequence code represents a job. The *i*th occurrence of the same value means the *i*th operation of this job. Here, the MOJSSP aims to find an order to optimize the above two objectives simultaneously. The constraints include each job should be processed on only one machine at a time, each machine can process only one job at a time, and the preemption is not allowed [33].

Notations used are listed below:

i, h: index of jobs, $i, h \in \{0, 1, 2, \ldots, n\}$

j, l: index of operations in a given job, $j, l \in \{0, 1, 2, \ldots, o\}$

k, g: index of machines, $k, g \in \{0, 1, 2, \ldots, m\}$

m: number of machines

n: number of jobs

o: number of operations in a given job

o_{ij}: the *j*th operation of the job i

C_i: the completion time of the job i

T_{ijk}: the processing time of the *j*th operation of the job i on machine M_k

E_{ijk} the completion time of the *j*th operation of the job i on machine M_k

D_i: the due date of the *i*th job

M_{ij}: set of available machines for the *j*th operation of job i

Index:

$$X_{ijk} = \left\{ \begin{array}{l} 1, \text{ if } o_{ij} \text{ is machined on } M_k \\ 0, \text{ otherwise} \end{array} \right\}$$

Two objective functions are simultaneously minimized:

$$\min \quad F1 = \max\{C_i | i = 0, 1, \ldots, n\} \tag{1}$$

$$F2 = \sum_{i=0}^{n} (C_i - D_i) \tag{2}$$

Subject to:

$$E_{ijk} > 0, \ i = 0, 1, \ldots, n; j = 0, 1, \ldots o; k = 0, 1, \ldots, m \tag{3}$$

$$E_{i(j+1)g} - E_{ijk} \geq X_{i(j+1)g} \times T_{i(j+1)g}, i = 0, 1, \ldots, n; j = 0, 1, \ldots, o - 1; k, g = 0, 1, \ldots, m \tag{4}$$

$$E_{ijk} - E_{hlk} \geq X_{ijk} \times T_{ijk} \ \text{or} \ E_{hlk} - E_{ijk} \geq X_{hlk} \times T_{hlk}, i = 0, 1, \ldots, n; j = 0, 1, \ldots, o; k = 0, 1, \ldots, m \tag{5}$$

$$\sum_{j=0}^{o} X_{ijk} = 1, \quad i = 0, 1, \ldots, n; k = 0, 1, \ldots, m \tag{6}$$

Equations (1) and (2) are used to minimize the makespan (F1) and total tardiness (F2), respectively. Equation (3) is the variable restriction. Equation (4) ensures the sequence of operations. Equation (5) is used to constrain operations so that they do not overlap. Equation (6) indicates an operation of a job could only be processed on one machine, that is, an operation cannot be divided, and it can only be processed on one machine from the beginning to completion.

4. The Hybrid Machine Learning and Population Knowledge Mining Method

The main process of the approach for MOJSSP (PAMOJSSP) is briefly illustrated in Figure 1. The process is started from the operations' attributes assignment. In this step, the attributes of each operation are deduced from the optimized objective functions, corresponding to the process information of each operation. These attributes' information is embedded into the individual class and then, under the calculation of metaheuristic algorithm, the Pareto frontier solutions presented as the optimal results among these non-dominated solutions are obtained. This series of Pareto frontier can be used as training data to obtain potential rules in these non-dominated solutions through data mining. After the acquisition of rules, an operation priority table is obtained by the rule correspondence and the ADSM. The rule-based initial population is eventually acquired by crossing the genes of the parent individual in the operation priority table. In the last step, the rule-based initial population combined with the hybrid metaheuristic algorithm obtains the new better Pareto frontier solutions for MOJSSP.

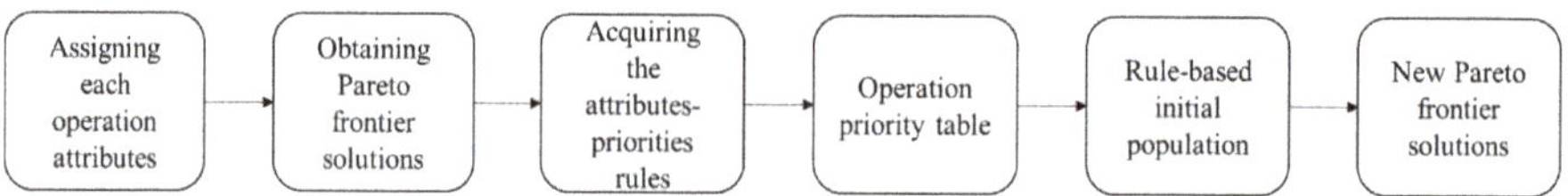

Figure 1. Process of the proposed approach for MOJSSP.

The following part introduces the concrete implementation of the proposed method.

4.1. Attributes

On the basis of attribute-oriented induction [34], combined with the optimization objectives of this problem, five attributes are finally determined: priority, operation feature, processing time, remaining time, and due date. Each attribute is classified into two or more types and explained in the following subsections.

4.1.1. Priorities

Priorities mean the position of each operation in the final sequence code. From the benchmark of 10×10 job shop problem, 100 positions could be acquired and needed to be sequenced to optimize the makespan and total tardiness. Ten classes of priorities are divided, and each priority has ten positions. That is, we divide the operation sequence code into 10 equal parts from front to back, with the priority of 0, 1, 2, … , 9, respectively. The number 0 indicates the highest priority level, while number 9 indicates the lowest priority level.

4.1.2. Operation Feature

Operation feature means the position of each operation in its route process. Referring to Nasiri et al. [30], the first operation is classified as "first". The second and third operations are labeled with "secondary". The fourth, fifth and sixth operations are classified as "middle". The seventh, eighth, and ninth operations mean "later". Finally, the tenth is labeled with "last".

4.1.3. Processing Time

Processing time is the required processing time for each operation on its predetermined machine. For LA18, we classified the processing time into three equal part as "short", "middle", and "long". As illustrated in Figure 2, the processing time with less than 37 is classified as "short". The time between 37 and 67 is "middle" and the left time is labeled with "long".

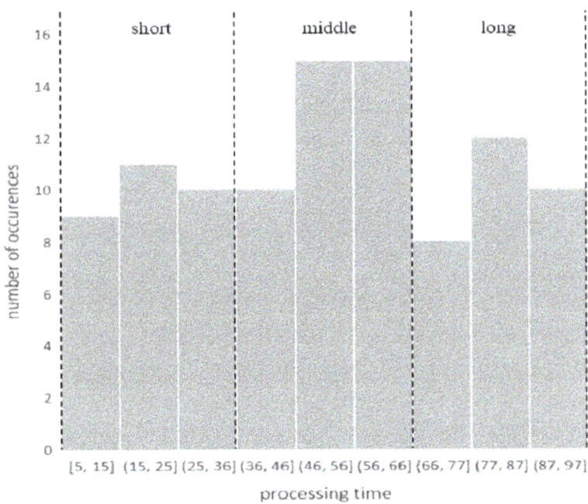

Figure 2. Processing time attribute.

4.1.4. Remaining Time

Remaining time refers to the accumulative processing time for the remaining operations to be processed after the current operation. As with the classification of processing time, its histogram is shown in Figure 3. The remaining time with less than 202 is defined as "short", the time between 202 and 402 is classified as "middle", and the time with more than 402 is labeled as "long".

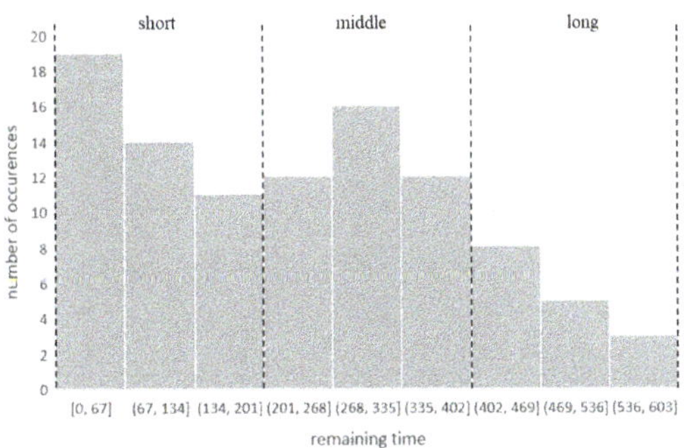

Figure 3. Remaining time attribute.

4.1.5. Due Date

Due date is the time when a job must be delivered. For optimizing the total tardiness, it is a critical attribute which needs to be considered as an important input of data mining. Because the benchmark LA18 only provides the processing time information, here we add the due date for each job. According to the benchmark, the shortest makespan is 848. Therefore, we could set the due date of Jobs 0–9 as a random number from 1100 to 1290 and arrange them in order. The due date for Job 0 is the most urgent and the due dates of the remaining jobs increase in order, which means that the due dates for jobs become looser. Table 1 shows the values of due date attribute for jobs.

Table 1. Due date attribute.

Job No.	Due Date	Class
0	1100	tight
1	1150	tight
2	1150	tight
3	1180	tight
4	1180	tight
5	1200	slack
6	1200	slack
7	1250	slack
8	1250	slack
9	1290	slack

4.2. Data Preparation

First, the operation feature, processing time, remaining time, and due date of each operation in LA18 are matched one by one using the above methods. Table 2 lists some of its sample data. The difference of attributes can be obtained by the significant test ($\alpha = 0.05$) [35]. The null hypothesis is that the significant difference of each column (each attribute) does not exist. The number of samples before matching is 100, while the p-value is 0.000, which is much less than 0.05. That is, the difference between columns is significant. After this, in the 9088 datasets obtained by 30 independent runs of NSGA-II combined with simulated annealing, 6645 non-dominated solutions were obtained after removing the repeated solutions. Among them, 493 sets of Pareto frontier solutions were obtained, and 66 sets of solutions were finally selected for the computational complexity, which forms 6600 transactions in a relational database. Table 3 shows the sample data of the trained data.

Table 2. Samples of partial data for operation attributes.

Operation	Machine	Processing Time	Remaining Time	Operation Feature	Processing Time	Remaining Time	Due Date
00	6	54	507	first	middle	long	tight
01	0	87	420	secondary	long	long	tight
02	4	48	372	secondary	middle	middle	tight
03	3	60	312	middle	middle	middle	tight
04	7	39	273	middle	middle	middle	tight
05	8	35	238	middle	short	middle	tight
...	...	...	...	...	...	...	...
97	9	85	105	later	long	short	slack
98	5	46	59	later	middle	short	slack
99	0	59	0	last	middle	short	slack

Table 3. Sample data of the trained data.

ID	Operation	Operation Feature	Processing Time	Remaining Time	Due Date	Priority
0	40	first	long	middle	tight	0
1	40	first	long	middle	tight	0
2	90	first	middle	long	slack	0
3	90	first	middle	long	slack	0
4	90	first	middle	long	slack	0
5	40	first	long	middle	tight	0
6	40	first	long	middle	tight	0
7	80	first	short	long	slack	0
8	80	first	short	long	slack	0
9	00	first	middle	long	tight	0
...	...	...	...	...	...	...
6594	99	last	middle	short	slack	9
6595	79	last	middle	Short	slack	9
6596	99	last	middle	short	slack	9
6597	89	last	short	short	slack	9
6598	69	last	long	short	slack	9
6599	69	last	long	short	slack	9

4.3. Rule Mining

Rule mining refers to mining association rules between attribute set {operation feature, processing time, remaining time, due date} and attribute {priority}. The decision tree is an effective means of mining classification. Here, we can get 33 rules from the theory of entropy, which is shown in Supplementary Materials File S1. However, the mined rules may have different priorities for the same attribute set. To discriminate this difference and enhance the richness of excellent populations, the weight of a priority class [12] in this paper is used to mine the rules behind the training data. The minimum value of the weight we set here is 0.01, which is the lower bound of the weight of the priority class. When the weight of a rule set under a priority class is less than this value, the rule set is considered not to belong to the priority class. The obtained 37 dominant rules are shown in Table 4.

The rules also could be expressed in the form of "if-then". For example, Rule 3 is given as:

If (operation feature = "first", and processing time = "middle", and remaining time = "long", and due date = "slack")

Then, (priority = "0", weight = "0.71"; priority = "1", weight = "0.28"; priority = "2", weight = "0.01")

It should be mentioned that there may be a case where the sum of weights is not equal to 1, which is due to rounding and does not affect the results of rules.

Table 4. Dominant rules obtained from knowledge mining.

Id	Rule Set	Priority									
		0	1	2	3	4	5	6	7	8	9
0	first long long slack	0.67	0.15	0.14	0.05						
1	first long middle tight	0.80	0.20								
2	first middle long slack	0.71	0.28	0.01							
3	first middle long tight	0.98	0.02								
4	first middle middle tight	0.82	0.18								
5	first short long slack	1.00									
6	first short long tight	0.33	0.48	0.12	0.07						
7	last long short slack						0.02	0.01	0.09	0.11	0.78
8	last long short tight						0.08	0.76	0.17	0.00	0.00
9	last middle short slack							0.07	0.02	0.17	0.75
10	last middle short tight							0.08	0.21	0.39	0.32
11	last short short slack								0.03	0.27	0.70
12	last short short tight						0.06	0.16	0.02	0.14	0.63
13	later long short slack				0.01	0.08	0.09	0.06	0.19	0.39	0.19

Table 4. *Cont.*

Id	Rule Set	Priority									
		0	1	2	3	4	5	6	7	8	9
14	later long short tight					0.12	0.13	0.16	0.29	0.13	0.18
15	later middle middle slack				0.05	0.26	0.05	0.26	0.39		
16	later middle short slack					0.12	0.18	0.27	0.24	0.19	
17	later middle short tight				0.01	0.06	0.08	0.17	0.13	0.43	0.12
18	later short short slack					0.05	0.13	0.12	0.18	0.37	0.15
19	later short short tight					0.18	0.53	0.10	0.15	0.04	
20	middle long middle slack		0.01	0.16	0.24	0.24	0.14	0.13	0.07		
21	middle long middle tight			0.14	0.37	0.04	0.08	0.36	0.02		
22	middle long short tight			0.05	0.27	0.40	0.21		0.08		
23	middle middle middle slack			0.01	0.19	0.22	0.13	0.17	0.14	0.11	0.02
24	middle middle middle tight		0.02	0.24	0.39	0.26	0.09				
25	middle middle short tight				0.30	0.36	0.15	0.18			
26	middle short middle slack			0.14	0.14	0.52	0.14	0.07			
27	middle short middle tight		0.07	0.18	0.15	0.11	0.08	0.22	0.17	0.02	
28	middle short short slack					0.18	0.26	0.03	0.48	0.05	
29	middle short short tight				0.36	0.12	0.30	0.06	0.15		
30	secondary long long slack	0.03	0.62	0.17	0.05	0.04	0.05	0.05	0.01		
31	secondary long long tight	0.03	0.82	0.15							
32	secondary middle long slack	0.38	0.39	0.22	0.02						
33	secondary middle middle slack		0.14	0.39	0.30	0.02	0.06	0.04	0.06		
34	secondary middle middle tight	0.11	0.23	0.34	0.21	0.08	0.03	0.01			
35	secondary short long slack	0.38	0.42	0.20							
36	secondary short middle tight	0.18	0.33	0.39	0.10						

4.4. Initial Population Generated Using ADSM

In this section, a rule-based initial population is finally obtained through the proposed ADSM. First, each operation obtains its possible position according to the mined rules. For example, from Table 2, attributes of Operation 00 are first, middle, long, and tight, which are consistent with the rule set with ID 3 in Table 4. It means that the priority weight of the first operation of the Job 0 is {priority = 0, weight = 0.98; priority = 1, weight = 0.02}. The highest priority level is initially defined as the possible position of the operation. Therefore, the initial priority of Operation 00 is 0. In Supplementary Materials File S2, these maximum values are labeled and all other priority weights can be obtained by traversing all operations. Next, the operations in each priority are readjusted to satisfy the requirement of ten operations per priority. To gain this purpose, ADSM is proposed to overcome problems that may arise during the sorting process.

As shown in Figure 4, we define the marked weight as the priority class of each operation. In the process, there are three main situations. The first situation is that the priority order of operations may be incorrect because operations are processed sequentially. That is, the weight of the marked class of the previous operation may be higher than the next operation. While confronting this problem, these two marked weights of the operations should be compared. Through calculating the difference between the marked weight of each operation and the weight of the current row corresponding to the column of the other marked class, the smaller one should remove the mark from the current weight to the new weight of the current row where the column corresponds to the other marked weight.

After adjusting the order of all operations, the next problem is to assign the operations between each adjacent priority group to meet the requirement of ten operations per priority. There are two situations, namely ">10" and "<10", which are the remaining two situations mentioned above. The second situation (the number of marked class in the current priority column >10) takes the deletion sorting method, that is removing the extra marked weights to the next priority column. The mark with the smallest difference is removed, which here refers to the difference between the weight of the last marked operation for each job in the current column and the adjacent weight in the next priority column. In the third situation (the number of marked class in the current priority column <10), the addition sorting method is used to add marked classes from the next priority column to the current priority column. The mark with the smallest difference is added first, and the standard is the difference between the weight of the first marked operation for each job in the next column and the corresponding weight in the current priority column. In the cases that the minimum difference is the same, the deletion sorting method first removes the marked operations with the higher encoded number, while the addition sorting method first provides the marked operations with the smaller encoded number.

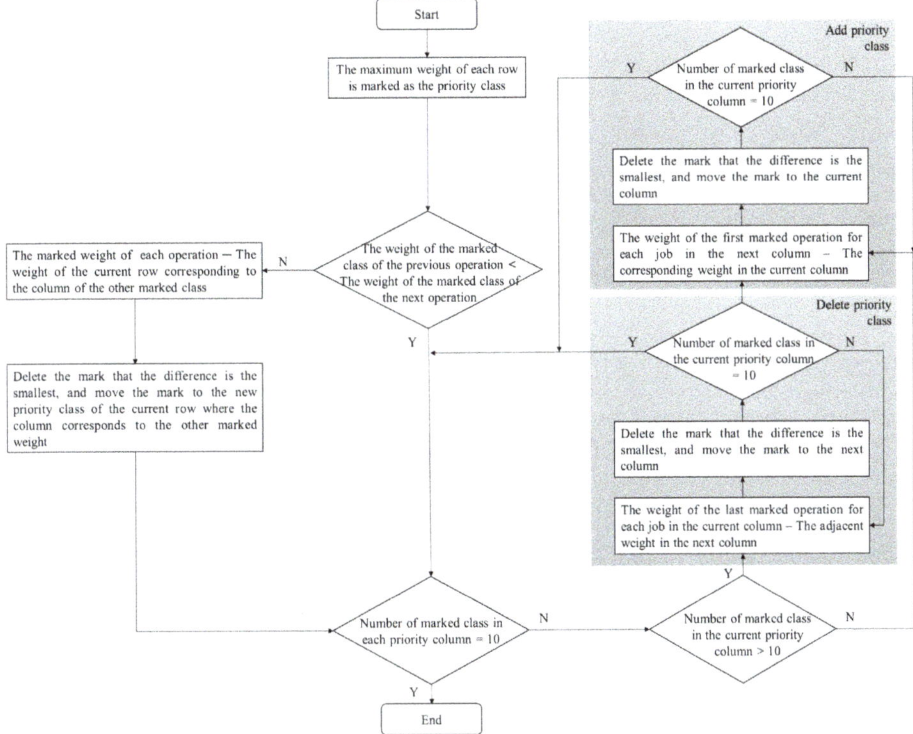

Figure 4. Framework of ADSM.

5. Experiment

The proposed method was applied to a well-known multi-objective metaheuristic algorithm, which runs 10 times per calculation in Eclipse. To better explain this method, a rule-based initial population and a mixed initial population mined by this method were compared with the random population under different criteria based on the benchmark LA18.

5.1. Selected Algorithm

For the multi-objective scheduling problem, the most popular algorithm is NSGA-II. However, it lacks a better local search capability. It means that, when solving some cases, its results usually fall into a locally optimal solution. To make experimental results more accurate and better, this study used the NSGA-II combined with simulated annealing (SA). The following introduces the NSGA-II and SA algorithm and the parameters used.

1. Nondominated sorting genetic algorithm II (NSGA-II)

     Based on NSGA, developed by Deb et al. [36], NSGA-II has the advantage of fast running speed, good robustness, and fast convergence. The encoding method is consistent with the above described. Each gene means an operation. Each chromosome/individual represents a solution or a sequence of all the operations waiting to be scheduled. NSGA-II combines parent and offspring into a new population and then obtains the non-dominated solutions by fast non-dominated sorting. The parent is generated by disrupting genes in the chromosome. The offspring is generated by the crossover and mutation method. The parameters are shown in Table 5.

2. Simulated annealing (SA)

Simulated annealing is an approximate method based on Monte Carlo design. It was first introduced by Kirkpatrick et al. [37] to solve the optimization problem. SA accepts a solution that is worse than the current solution by the Metropolis criterion, thus it is possible to jump out of this local optimal solution to reach the global optimal solution. Table 6 lists the parameters used in SA.

Table 5. Parameters used in NSGA-II.

Parameters	Values
population size	25, 50, 100
mutation rate	0.002
crossover rate	0.9
size of tournament selection	10
number of iterations	100, 300

Table 6. Parameters used in SA.

Parameters	Values
population size	25, 50, 100
initial temperature	100
end temperature	0.01
cooling rate	0.001
number of iterations	100, 300

The non-dominated solutions were obtained by NSGA-II, and then SA was used to local search. Because of the multi-objective reasons, an objective function was randomly selected as the direction of a search, and the obtained non-dominated solutions were stored in the external archive. SA was applied to local search every 50 generations in this experiment.

To obtain better initial population rules, the data source of data mining must be the optimal or near-optimal solution set, so that accurate and reliable rules can be discovered. In Tables 7 and 8, we can see that the data obtained by NSGA-II hybrid with SA are more effective, thus the results obtained by this algorithm were selected as the data source of knowledge mining. More comparisons between this algorithm and other algorithms can be seen in [38].

To verify the knowledge mining method, three different initial populations of NSGA-II combined with simulated annealing were considered as follows:

- Knowledge mining heuristic optimization method (rule)
  The initial population was generated using the proposed hybrid machine learning and population knowledge mining method.
- Heuristic optimization method (random)
  The initial population of this method was completely randomized.
- Hybrid population optimization method (mixed)
  Half of the initial population was generated by knowledge mining and the other half was randomly generated.

Table 7. Makespan of the NSGA-II and NSGA-II + SA.

IP	IT		NSGA-II	NSGA-II + SA
25	100	best	933	853
		average	974	909
	300	best	895	848
		average	947	910
50	100	best	895	854
		average	941	912
	300	best	865	848
		average	931	915
100	100	best	898	848
		average	938	919
	300	best	861	848
		average	912	943

Table 8. Total tardiness of the NSGA-II and NSGA-II + SA.

IP	IT		NSGA-II	NSGA-II + SA
25	100	best	−3860	−4529
		average	−3350	−4338
	300	best	−4001	−4524
		average	−3757	−4537
50	100	best	−4241	−4630
		average	−3743	−4400
	300	best	−4279	−4815
		average	−3982	−4569
100	100	best	−4273	−4698
		average	−3703	−4495
	300	best	−4539	−4698
		average	−4229	−4587

5.2. Performance Metrics

Three popular metrics were employed to evaluate the performance of the proposed method for MOJSSP: Relative Error (RE), Coverage of two sets (Cov) [39], and Spacing [40]. They can be expressed as follows:

1. Relative Error (RE)

   Extending the method of Arroyo and Leung [41], we analyzed the performance of the two acquired objectives using the relative error (RE) metric. The formulation is as follows:

$$RE = \frac{\overline{F} - F_{\text{best}}}{F_{\text{best}}} \times 100 \tag{7}$$

where $\overline{F}$ is the mean value of makespan or total tardiness and F_{best} is the best makespan or total tardiness obtained among the iterations.

2. Coverage of Two Sets (Cov)

   This indicator was used to measure the dominance between two sets of solutions. The definition is as follows:

$$Cov(X, Y) = \frac{\left|\{y \in Y; \exists x \in X : x <= y\}\right|}{|Y|} \tag{8}$$

where X and Y are two non-dominated sets to be compared. x and y are subsets of X and Y, respectively. "$x <= y$" represents x dominates y or $x = y$. The value $Cov(X, Y) = 1$ means that all solutions in Y are dominated by or equal to solutions in X. The opposite, $Cov(X, Y) = 0$ represents that no solutions in Y are dominated by the set X. It should be mentioned that there is not necessarily a relationship between $Cov(X, Y)$ and $Cov(Y, X)$. Therefore, these two values should be calculated independently. $Cov(X, Y) > Cov(Y, X)$ means the set X is better than the set Y.

3. Spacing

It measures the standard deviation of the minimum distance from each solution to other solutions. The smaller the Spacing value is, the more uniform the solution set is. The expression is as follows:

$$Spacing = \sqrt{\frac{1}{n-1}\sum_{1}^{n}\left(\bar{d} - d_i\right)^2} \tag{9}$$

where n is the number of solutions in the obtained Pareto frontier, d_i is the minimum distance between the solution and its nearest solution, and $\bar{d}$ is the average value of all d_i.

5.3. Results and Discussion

In this section, we first compare the new rule method, mixed method, and traditional random method under different iteration times with the same initial population size. Then, under the same number of iterations and different initial population sizes, the performance of the three methods are compared. Finally, compared with Nasiri's rule-based initial population [30], the effectiveness of the ADSM is proved.

To analyze the values of the two objective functions and the average distribution from the optimal solution, the result of different iterations of 100 initial population is shown in Table 9, where F1 and F2 represent the values of makespan and total tardiness, respectively, and the bold contents are better values. The rules method represents the new approach for population knowledge mining applied to the hybrid NSGA-II to solve MOJSSP. The traditional random method is the NSGA-II combined with SA for MOJSSP. The mixed method is a combination of the two methods. Figure 5 shows the box distribution of its function values of which the values in the rule method and the mixed method are overall smaller than the traditional method. To some extent, the smaller are the values, the better is the performance. Similarly, the results of different iterations of 50 and 25 initial population are shown in Tables 10 and 11. Figures 6 and 7 are the box diagrams of the respective function values. From the results, we can find that the population mining method has more excellent results than the traditional random method because the rule population contains more excellent information than the traditional random population. Although the new method has no absolute advantage in the test of 25 initial population, it should be explained that, when the initial population size is too small, the performance of the method will inevitably be reduced. In general, there is no obvious advantage or disadvantage between rule method and mixed method in terms of function values, but, in term of the index RE, the rule method is superior to the mixed method, and both are superior to the random method. That is, the more rule-based individuals there are in the population, the more excellent information they contain, and thus it is easier to obtain excellent solutions under a limited number of iterations.

Table 9. Values of makespan, total tardiness, and RE of 100 initial population.

Methods	Iteration		F1	F2	RE
rules	100	best	**848**	−4696	7.5|4
		average	**912**	**−4500**	/
	300	best	**848**	−4717	**6.7|2**
		average	**905**	−4622	/
mixed	100	best	848	−4618	8.1|**2.7**
		average	916.3	−4492	/
	300	best	848	**−4846**	10.7|4.3
		average	939	**−4639**	/
random	100	best	848	**−4698**	8.4|4.3
		average	919	−4495	/
	300	best	848	−4698	11.2|2.4
		average	940	−4587	/

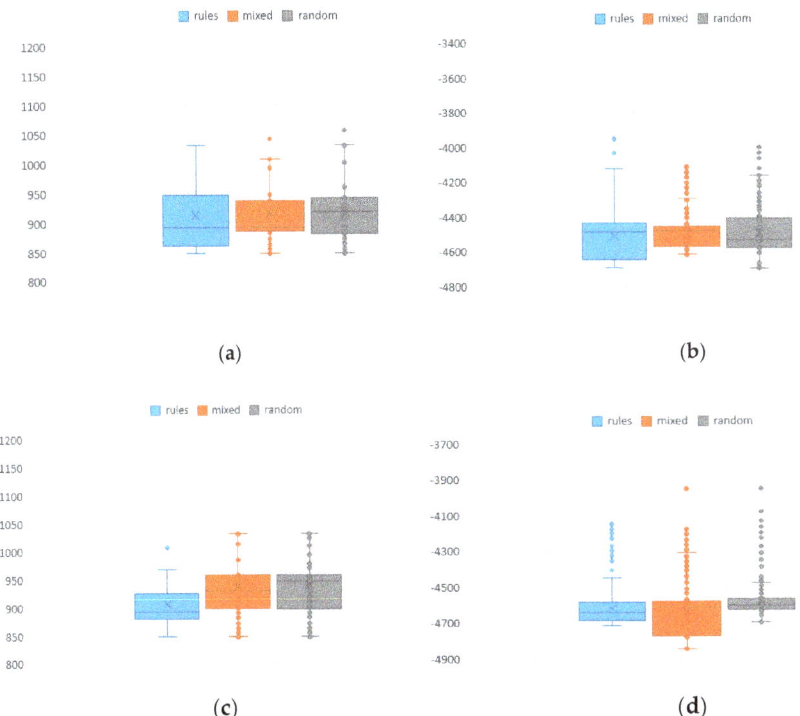

Figure 5. Distribution of makespan and total tardiness of 100 initial population sizes under different iteration numbers: (**a**) makespan under 100 iterations; (**b**) total tardiness under 100 iterations; (**c**) makespan under 300 iterations; and (**d**) total tardiness under 300 iterations.

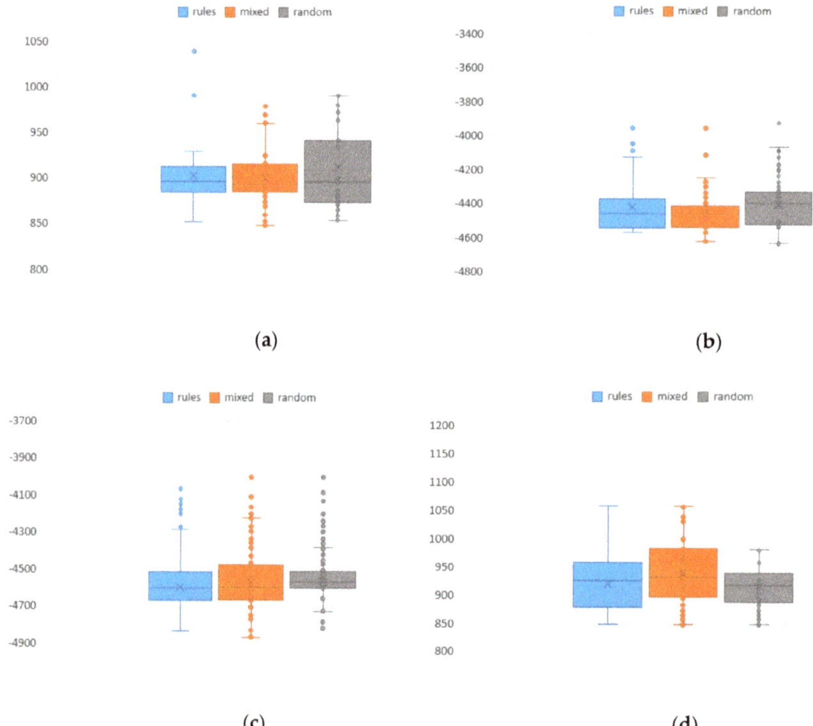

(a)

(b)

(c)

(d)

Figure 6. Distribution of makespan and total tardiness of 50 initial population sizes under different iteration numbers: (**a**) makespan under 100 iterations; (**b**) total tardiness under 100 iterations; (**c**) makespan under 300 iterations; and (**d**) total tardiness under 300 iterations.

Table 10. Values of makespan, total tardiness, and RE of 50 initial population.

Methods	Iteration		F1	F2	RE
rules	100	best	852	−4568	**5.94|3.27**
		average	902.6	−4418.4	/
	300	best	**848**	−4835	8.54|**4.87**
		average	920.4	**−4599.3**	/
mixed	100	best	**848**	−4618	6.03|3.52
		average	**899.1**	**−4455.6**	/
	300	best	848	**−4868**	10.74|6.04
		average	939.1	−4573.8	/
random	100	best	854	**−4630**	6.77|4.97
		average	911.8	−4400	/
	300	best	848	−4815	**7.88**|5.12
		average	**914.8**	−4568.5	/

Table 11. Values of makespan, total tardiness, and RE of 25 initial population.

Methods	Iteration		F1	F2	RE
rules	100	best	**852**	**−4693**	8.49|6.58
		average	924.3	−4384.3	/
	300	best	**848**	−4731	**5.83|4.34**
		average	**897.4**	−4525.9	/
mixed	100	best	855	−4593	5.38|5.11
		average	**901**	−4358.2	/
	300	best	848	−4810	7.63|5.59
		average	912.7	**−4540.9**	/
random	100	best	853	−4529	6.58|**4.22**
		average	909.1	−4337.8	/
	300	best	848	**−4824**	7.28|5.94
		average	909.7	−4537.4	/

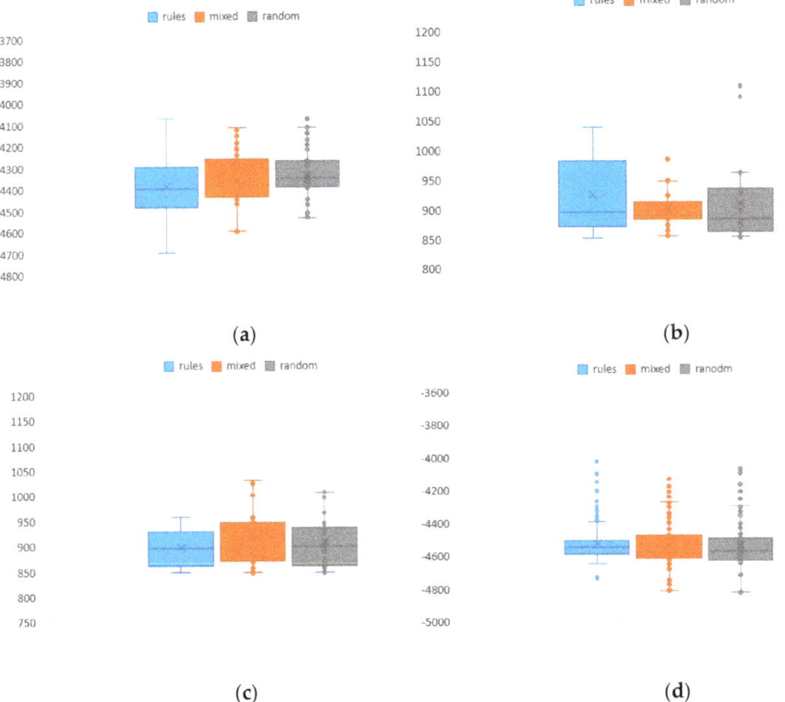

Figure 7. Distribution of makespan and total tardiness of 25 initial population sizes under different iteration numbers: (**a**) makespan under 100 iterations; (**b**) total tardiness under 100 iterations; (**c**) makespan under 300 iterations; and (**d**) total tardiness under 300 iterations.

Since this paper optimizes two objectives at the same time, the values of makespan or total tardiness cannot be separately compared. Thus, the efficiency of the method was determined by the performance metrics of the acquired Pareto solutions, especially the dominance metric (Cov). Here, the performance metrics of the three methods are compared under different initial population sizes with the same number of iterations, as shown in Tables 12–14, where IP represents the size of the initial population and the bold contents are the better values. In Table 12, the Cov value of the rule method is generally higher than that of the traditional method. The larger is the Cov value, the higher is the dominant ability. That is, in terms of dominance, the rule method is superior to the traditional random method in comparing the size of different initial populations. At the same time, the Spacing values of the new method is generally smaller than that of the traditional method. From Equation (9), it can be seen that the distribution of the rule method is also superior to the random method in considering the initial population size as a variable. Figure 8 shows the box diagram of the Cov of the rule method vs. the random method. From the box diagram, we can find more intuitively that the Pareto solution of the rule method is superior to that of the traditional random method. Similarly, from the results in Tables 13 and 14, we can conclude that the performance of rule method in Cov is also "rules > mixed > random". For Spacing performance, the mixed method is better than the rule method, which shows that the mixed method can more easily obtain a uniformly distributed solution set with a limited number of iterations. Overall, the Spacing metrics of both the rule and mixed methods are superior to the random method.

Table 12. Values of Cov and Spacing (rules vs. random).

Methods	IP		IT100		IT300	
			Cov	Spacing	Cov	Spacing
rules	25	best	1	37	1	8.8
		average	0.54		0.4	
	50	best	1	25	1	39
		average	0.59		0.38	
	100	best	1	29	1	22
		average	0.56		0.47	
random	25	best	0.75	76	1	50.5
		average	0.19		0.35	
	50	best	0.9	65	1	45
		average	0.21		0.43	
	100	best	0.8	37	0.86	18
		average	0.19		0.29	

Table 13. Values of Cov and Spacing (mixed vs. random).

Methods	IP		IT100		IT300	
			Cov	Spacing	Cov	Spacing
mixed	25	best	1	19	1	48
		average	0.63		0.31	
	50	best	1	24	0.86	37
		average	0.45		0.35	
	100	best	1	16	0.92	20
		average	0.41		0.45	
random	25	best	0.27	76	1	50.5
		average	0.11		0.43	
	50	best	1	65	1	45
		average	0.33		0.4	
	100	best	1	37	1	18
		average	0.39		0.33	

Table 14. Values of Cov and Spacing (rules vs. mixed).

Methods	IP		IT100		IT300	
			Cov	**Spacing**	**Cov**	**Spacing**
rules	25	best	**1**	37	**1**	**8.8**
		average	0.24		**0.46**	
	50	best	**1**	25	**1**	39
		average	**0.47**		0.44	
	100	best	**1**	29	**1**	22
		average	**0.53**		**0.41**	
mixed	25	best	1	**19**	0.8	48
		average	**0.41**		0.24	
	50	best	1	**24**	0.88	37
		average	0.31		**0.41**	
	100	best	1	**16**	1	**20**
		average	0.21		0.37	

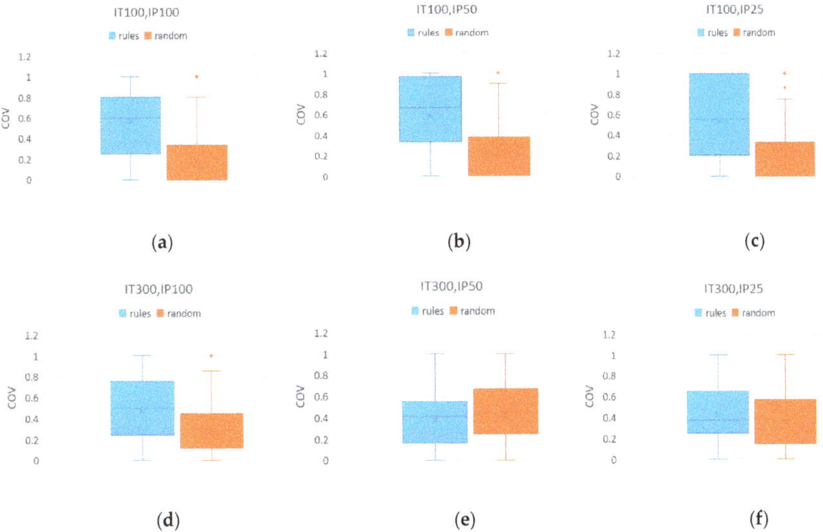

Figure 8. Box diagram of Cov of the rule method vs. random method with different sizes of initial population under 100 and 300 iterations: (**a**) IP = 25 under 100 iterations; (**b**) IP = 50 under 100 iterations; (**c**) IP = 100 under 100 iterations; (**d**) IP = 25 under 300 iterations; (**e**) IP = 50 under 300 iterations; and (**f**) IP = 100 under 300 iterations.

To better illustrate the proposed ADSM, here we compare it with the Nasiri's method. Compared with the operation sequence and knowledge constraints that are not considered, Table 15 shows that the proposed ADSM has an impact on the value of the single objective function. Although the advantage of the ADSM is not obvious in terms of single objective value, this is because the influence of local adjustment of operations on a single objective value is relatively small under multiple iterations. From the perspective of multi-objective, it significantly improves the performance metrics of dominance and distribution, as shown in Table 16. The results show that, under different iterations and initial population sizes, the Cov values and Spacing values of the ADSM are better, which demonstrates that it is more effective in solving multi-objective problems. On the other hand, the validity of the method

ADSM is proved, which shows that the new proposed method can overcome the problem of sequencing the operations under knowledge mining as well as meet the requirement of high performance.

Table 15. Comparison of function values of two rule-based populations.

Methods	IP	IT		F1	F2	RE
ADSM	25	100	best	**852**	**−4693**	8.49|6.58
			average	924.3	**−4384.3**	/
		300	best	**848**	−4731	**5.83|4.34**
			average	**897.4**	−4525.9	/
	50	100	best	852	**−4568**	5.94|3.27
			average	**902.6**	−4418.4	/
		300	best	**848**	**−4835**	8.54|**4.87**
			average	920.4	**−4599.3**	/
	100	100	best	**848**	**−4696**	7.5|4
			average	912	**−4500**	/
		300	best	**848**	−4717	**6.7|2**
			average	**905**	**−4622**	/
Nasiri's	25	100	best	853	−4587	4|**5.8**
			average	**887**	−4319	/
		300	best	848	**−4810**	8.84|5.51
			average	923	**−4545**	/
	50	100	best	**848**	−4558	7.8|**1.34**
			average	914.2	**−4497**	/
		300	best	848	−4822	7.1|5.4
			average	**908**	−4561.5	/
	100	100	best	848	−4642	**6|3.8**
			average	**896.3**	−4467	/
		300	best	848	**−4799**	12.7|4.6
			average	956	−4576	/

Table 16. Comparison of Cov and Spacing of two rule-based populations.

Methods	IP		IT100		IT300	
			Cov	Spacing	Cov	Spacing
ADSM	25	best	**1**	37	**1**	8.8
		average	**0.54**		**0.42**	
	50	best	**1**	25	**1**	39
		average	**0.59**		**0.47**	
	100	best	**1**	29	**1**	22
		average	**0.55**		**0.39**	
Nasiri's	25	best	0.75	41	1	44
		average	0.19		0.35	
	50	best	0.9	46	1	29
		average	0.21		0.33	
	100	best	0.8	71	1	64
		average	0.16		0.37	

6. Conclusions

This paper presents a population knowledge mining approach combined with a hybrid machine learning algorithm to solve the MOJSSP with makespan and total tardiness criteria. The purpose of this method is to find a rule-based initial population from good production planning, which can produce better Pareto solutions than the traditional stochastic methods. Many optimal or near-optimal solutions are created by the method of NSGA-II integrated with SA on a benchmark of 10×10 job shop problem (LA18), and a training dataset with selected operations' attributes is therefore acquired. For mining knowledge, the weight of priority is used to extract 37 potential rules, which map the relationship between the attribute set and priority.

To form the rule-based initial population, a complicated ADSM is proposed to assign each operation a priority class. This sorting method overcomes the problem of misalignment of the sequence of operations and overcomes the problem of insufficient or overflowing operations owned by the priority class. After generating an initial population set based on the data mining and ADSM, we compared it with the NSGA-II hybrid SA method of the mixed, random, and Nasiri's initial population. Considering the different numbers of iterations and the sizes of the initial population, a series of comparative and computational experiments were conducted. The results show that the new proposed method is better than the traditional method in terms of the magnitude and relative error of each objective function, and the Cov and Spacing performance index of Pareto solutions.

The future works of our study will mainly focus on other methods of initial population formation and the application of the proposed method to other cases. Besides, considering other multi-objective problems and how to choose a mining algorithm to better improve the effect of rules need further research.

Supplementary Materials: The following are available online at http://www.mdpi.com/2076-3417/9/24/5286/s1.

Author Contributions: Conceptualization, Y.Q.; data curation, Y.Q.; methodology, Y.Q.; project administration, W.J.; investigation, C.Z.; writing—original draft preparation, Y.Q.; writing—review and editing, Y.Q.

Funding: This research was funded by the Natural Science Foundation of Jiangsu Province, grant number BK20170190.

Conflicts of Interest: The authors declare no conflict of interest.

References

1. Duan, Y.; Edwards, J.S.; Dwivedi, Y.K. Artificial intelligence for decision making in the era of Big Data–evolution, challenges and research agenda. *Int. J. Inf. Manag.* **2019**, *48*, 63–71. [CrossRef]
2. Dai, H.; Wang, H.; Xu, G.; Wan, J.; Imran, M. Big data analytics for manufacturing internet of things: Opportunities, challenges and enabling technologies. *Enterp. Inf. Syst.* **2019**, 1–25. [CrossRef]
3. Kuo, Y.; Kusiak, A. From data to big data in production research: The past and future trends. *Int. J. Prod. Res.* **2019**, *57*, 4828–4853. [CrossRef]
4. Lin, S.; Dong, C.; Chen, M.; Zhang, F.; Chen, J. Summary of new group intelligent optimization algorithms. *Comput. Eng. Appl.* **2018**, *54*, 1–9.
5. Gao, L.; Pan, Q. A shuffled multi-swarm micro-migrating birds optimizer for a multi-resource-constrained flexible job shop scheduling problem. *Inf. Sci.* **2016**, *372*, 655–676. [CrossRef]
6. Ruiz, R.; Pan, Q.; Naderi, B. Iterated Greedy methods for the distributed permutation flowshop scheduling problem. *Omega* **2019**, *83*, 213–222. [CrossRef]
7. Dao, T.; Pan, T.; Nguyen, T.; Pan, J. Parallel bat algorithm for optimizing makespan in job shop scheduling problems. *J. Intell. Manuf.* **2018**, *29*, 451–462. [CrossRef]
8. Fan, B.; Yang, W.; Zhang, Z. Solving the two-stage hybrid flow shop scheduling problem based on mutant firefly algorithm. *J. Ambient Intell. Humaniz. Comput.* **2019**, *10*, 979–990. [CrossRef]
9. Li, M.; Lei, D.; Cai, J. Two-level imperialist competitive algorithm for energy-efficient hybrid flow shop scheduling problem with relative importance of objectives. *Swarm Evol. Comput.* **2019**, *49*, 34–43. [CrossRef]
10. Gong, G.; Deng, Q.; Chiong, R.; Gong, X.; Huang, H. An effective memetic algorithm for multi-objective job-shop scheduling. *Knowl. Based Syst.* **2019**, *182*, 104840. [CrossRef]
11. Wiemer, H.; Drowatzky, L.; Ihlenfeldt, S. Data mining methodology for engineering applications (DMME)—A holistic extension to the CRISP-DM model. *Appl. Sci.* **2019**, *9*, 2407. [CrossRef]

12. Koonce, D.A.; Tsai, S.C. Using data mining to find patterns in genetic algorithm solutions to a job shop schedule. *Comput. Ind. Eng.* **2000**, *38*, 361–374. [CrossRef]
13. Wu, B.; Hao, K.; Cai, X.; Wang, T. An integrated algorithm for multi-agent fault-tolerant scheduling based on MOEA. *Future Gener. Comput. Syst.* **2019**, *94*, 51–61. [CrossRef]
14. Huang, X. A hybrid genetic algorithm for multi-objective flexible job shop scheduling problem considering transportation time. *Int. J. Intell. Comput. Cybern.* **2019**, *12*, 154–174. [CrossRef]
15. Souier, M.; Dahane, M.; Maliki, F. An NSGA-II-based multiobjective approach for real-time routing selection in a flexible manufacturing system under uncertainty and reliability constraints. *Int. J. Adv. Manuf. Technol.* **2019**, *100*, 2813–2829. [CrossRef]
16. Ahmadi, E.; Zandieh, M.; Farrokh, M.; Emami, S.M. A multi objective optimization approach for flexible job shop scheduling problem under random machine breakdown by evolutionary algorithms. *Comput. Oper. Res.* **2016**, *73*, 56–66. [CrossRef]
17. Zhou, Y.; Yang, J.; Zheng, L. Hyper-heuristic coevolution of machine assignment and job sequencing rules for multi-objective dynamic flexible job shop scheduling. *IEEE Access* **2019**, *7*, 68–88. [CrossRef]
18. Zhang, R.; Chiong, R. Solving the energy-efficient job shop scheduling problem: A multi-objective genetic algorithm with enhanced local search for minimizing the total weighted tardiness and total energy consumption. *J. Clean. Prod.* **2016**, *112*, 3361–3375. [CrossRef]
19. Du, W.; Zhong, W.; Tang, Y.; Du, W.; Jin, Y. High-dimensional robust multi-objective optimization for order scheduling: A decision variable classification approach. *IEEE Trans. Ind. Inform.* **2019**, *15*, 293–304. [CrossRef]
20. Sheikhalishahi, M.; Eskandari, N.; Mashayekhi, A.; Azadeh, A. Multi-objective open shop scheduling by considering human error and preventive maintenance. *Appl. Math. Model.* **2019**, *67*, 573–587. [CrossRef]
21. Lu, C.; Gao, L.; Pan, Q.; Li, X.; Zheng, J. A multi-objective cellular grey wolf optimizer for hybrid flowshop scheduling problem considering noise pollution. *Appl. Soft Comput.* **2019**, *75*, 728–749. [CrossRef]
22. Qin, H.; Fan, P.; Tang, H.; Huang, P.; Fang, B.; Pan, S. An effective hybrid discrete grey wolf optimizer for the casting production scheduling problem with multi-objective and multi-constraint. *Comput. Ind. Eng.* **2019**, *128*, 458–476. [CrossRef]
23. Zhou, Y.; Yang, J.; Zheng, L. Multi-agent based hyper-heuristics for multi-objective flexible job shop scheduling: A case study in an aero-engine blade manufacturing plant. *IEEE Access* **2019**, *7*, 21147–21176. [CrossRef]
24. Fu, Y.; Wang, H.; Huang, M. Integrated scheduling for a distributed manufacturing system: A stochastic multi-objective model. *Enterp. Inf. Syst.* **2019**, *13*, 557–573. [CrossRef]
25. Ingimundardottir, H.; Runarsson, T.P. Discovering dispatching rules from data using imitation learning: A case study for the job-shop problem. *J. Sched.* **2018**, *21*, 413–428. [CrossRef]
26. Li, X.; Olafsson, S. Discovering dispatching rules using data mining. *J. Sched.* **2005**, *8*, 515–527. [CrossRef]
27. Olafsson, S.; Li, X. Learning effective new single machine dispatching rules from optimal scheduling data. *Int. J. Prod. Econ.* **2010**, *128*, 118–126. [CrossRef]
28. Jun, S.; Lee, S.; Chun, H. Learning dispatching rules using random forest in flexible job shop scheduling problems. *Int. J. Prod. Res.* **2019**, 3290–3310. [CrossRef]
29. Kumar, S.; Rao, C.S.P. Application of ant colony, genetic algorithm and data mining-based techniques for scheduling. *Robot. Comput. Integr. Manuf.* **2009**, *25*, 901–908. [CrossRef]
30. Nasiri, M.M.; Salesi, S.; Rahbari, A.; Salmanzadeh Meydani, N.; Abdollai, M. A data mining approach for population-based methods to solve the JSSP. *Soft Comput.* **2019**, *23*, 11107–11122. [CrossRef]
31. Hillier, F.S.; Lieberman, G.J. *Introduction to Operations Research*, 7th ed.; McGraw-Hill Science: New York, NY, USA, 2001.
32. Lawrence, S. *An Experimental Investigation of Heuristic Scheduling Techniques*; Carnegie-Mellon University: Pittsburgh, PA, USA, 1984.
33. Fu, Y.; Wang, H.; Tian, G.; Li, Z.; Hu, H. Two-agent stochastic flow shop deteriorating scheduling via a hybrid multi-objective evolutionary algorithm. *J. Intell. Manuf.* **2019**, *30*, 2257–2272. [CrossRef]
34. Cai, Y. Attribute-Oriented Induction in Relational Databases, Knowledge Discovery in Databases. Ph.D. Thesis, Simon Fraser University, Burnaby, BC, Canada, 1989.
35. González-Briones, A.; Prieto, J.; De La Prieta, F.; Herrera-Viedma, E.; Corchado, J. Energy optimization using a case-based reasoning strategy. *Sensors* **2018**, *18*, 865. [CrossRef]

36. Deb, K.; Pratap, A.; Agarwal, S.; Meyarivan, T. A fast and elitist multiobjective genetic algorithm: NSGA-II. *IEEE Trans. Evol. Comput.* **2002**, *2*, 182–197. [CrossRef]

37. Kirkpatrick, S.; Gelatt, C.D.; Vecchi, M.P. Optimization by simulated annealing. *Science* **1983**, *4598*, 671–680. [CrossRef] [PubMed]

38. Cobos, C.; Erazo, C.; Luna, J.; Mendoza, M.; Gaviria, C.; Arteaga, C.; Paz, A. Multi-objective memetic algorithm based on NSGA-II and simulated annealing for calibrating CORSIM micro-simulation models of vehicular traffic flow. In *Conference of the Spanish Association for Artificial Intelligence*; Springer: Cham, Switzerland, 2016.

39. Zitzler, E.; Thiele, L. Multiobjective evolutionary algorithms: A comparative case study and the strength pareto approach. *IEEE Trans. Evol. Comput.* **1999**, *3*, 257–271. [CrossRef]

40. Schott, J.R. *Fault Tolerant Design Using Single and Multicriteria Genetic Algorithm Optimization*; Massachusetts Institute of Technology: Cambridge, MA, USA, 1995.

41. Arroyo, J.E.C.; Leung, J.Y.T. An effective iterated greedy algorithm for scheduling unrelated parallel batch machines with non-identical capacities and unequal ready times. *Comput. Ind. Eng.* **2017**, *105*, 84–100. [CrossRef]

Article

Design and Simulation of a Capacity Management Model Using a Digital Twin Approach Based on the Viable System Model: Case Study of an Automotive Plant

Sergio Gallego-García [1,*], **Jan Reschke** [2,*] **and Manuel García-García** [1,*]

[1] Department of Construction and Fabrication Engineering, National Distance Education University (UNED), 28040 Madrid, Spain

[2] Institute for Industrial Management (FIR), RWTH Aachen University, 52074 Aachen, Germany

* Correspondence: sgallego118@alumno.uned.es (S.G.-G.); Jan.Reschke@fir.rwth-aachen.de (J.R.); mggarcia@ind.uned.es (M.G.-G.); Tel.: +34-682-880-591 (M.G.-G.)

Received: 21 October 2019; Accepted: 11 December 2019; Published: 17 December 2019

Abstract: Matching supply capacity and customer demand is challenging for companies. Practitioners often fail due to a lack of information or delays in the decision-making process. Moreover, researchers fail to holistically consider demand patterns and their dynamics over time. Thus, the aim of this study is to propose a holistic approach for manufacturing organizations to change or manage their capacity. The viable system model was applied in this study. The focus of the research is the clustering of manufacturing and assembly companies. The goal of the developed capacity management model is to be able to react to all potential demand scenarios by making decisions regarding labor and correct investments and in the right moment based on the needed information. To ensure this, demand data series are analyzed enabling autonomous decision-making. In conclusion, the proposed approach enables companies to have internal mechanisms to increase their adaptability and reactivity to customer demands. In order to prove the conceptual model, a simulation of an automotive plant case study was performed, comparing it to classical approaches.

Keywords: capacity planning; digital twin; forecasting; viable system model; simulation; system dynamics; customer demand; demand planning; automotive

1. Introduction

In a fast-changing environment with increasing demand volatility, shorter product life cycles, new technologies, and new trade regulations, balancing supply capacity and market demand is a challenge for organizations all over the world. In this context, the capability of an organization to change as an adaptability characteristic is key in order to secure its viability. As a consequence, global managers are under increasing pressure to make decisions more quickly.

While new technologies open new possibilities, many managers, even in well-established companies, do not have a system or virtual digital twin to enable them to foresee the expected outcomes of decisions that are to be made. Neither complete information nor all the implications of a decision are appropriately considered in most organizations. Because of this, many managers usually decide to start a process of collecting information to conduct an analysis as the basis for decision-making regarding capacity. This process involves different areas. Therefore, the decision is delayed until it has to be made, implying the loss of profit generation, as customers are not buying products due to a lack of capacity or long delivery times as a result.

For all of these reasons, this paper attempts to provide an answer to those managers and their daily planning challenges by providing a conceptual model and a simulation of a case study. As a consequence, the practical relevance of this paper is a digital interface for decision-making support including technical, management, and economic key performance indicators. This interface provides a platform that can be used to avoid unnecessary investments, increase sales for all types of manufacturing companies, reduce operating costs, and support higher service levels. Moreover, the theoretical relevance of the paper is a novel approach that combines already consolidated scientific methods into a new concept for demand and capacity planning that is oriented to end-customer demand.

The purpose of this paper is to develop a generic conceptual model for capacity planning in which decisions can be sped up in order to adapt them as quickly as possible to demands considering a level of risk for investments and changes in the production system. That is why economic parameters are considered to balance potential profit generation with investments or costs incurred. In this regard, the concept development presents different levels: network planning, location-related planning, and line planning within a location. With these three levels it can be taken into account whether it is a greenfield or brownfield scenario and whether it involves a new product launch.

The main hypothesis is that a manufacturing company capable of adapting its capacity to demand using the structure of the viable system model (VSM) will be able to react faster and, therefore, applying this model will have a positive impact on the achievement of short-, medium-, and long-term goals of production companies. The VSM approach increases the adaptability of a company to face future potential scenarios, because the company will be able to make strategic decisions that will later influence the tactical and operative levels.

Based on the conceptual model, a case study for an original equipment manufacturer (OEM) was chosen. It is a brownfield scenario in which there is a new product launch in a production line with two models using the same platform. By proving different scenarios and decisions to be made, the conceptual model is proven in order to check the hypothesis previously set.

2. Fundamental Definitions and State of Research

The literature review shows that there are about 30 basic methods that help to predict future sales, many of them with subtypes [1]. The purpose of demand planning is to improve decisions that affect demand accuracy [2]. Its main task is to calculate future demand, and it comprises the selection of a specific forecasting method and its parameters. The typical demand patterns are stationary, seasonal, trend, and sporadic [3].

Independent of the demand type, the bullwhip effect is a challenge in every supply chain due to the lack of transparency between stages. As a result, it is frequently difficult to forecast the final customer demand based on orders along the supply chain [4], and this causes an extreme fluctuation of stock at the beginning of the supply chain, creating many backorders alternating with large excess inventory [5].

To combat the bullwhip effect, there should be information sharing of consumer demand across the supply chain in order to allow each party to plan efficiently according to true demand information [6].

Collaborative forecasting and planning are the basis for capacity management through the supply chain. Capacity utilization, like stock levels and work in progress, is one of the degrees of freedom of planning and control in supply chains. An assessment of the necessary capacitive resources is a task in every planning period. Supply flexibility in medium- and short-term planning often requires long-term arrangements [5] such as investments in production capacity.

To decide on investments properly, it is necessary to focus on the system limitations. A constraint can be the management of an organization, capacities along the supply chain, or market demand. In this context, the theory of constraints (TOC) provides a framework to deal with capacity management issues with five steps: identify the constraint, exploit the constraint, subordinate the rest of the system to the constraint before increasing capacity, then add new capacities and start other times by finding the capacity constraint [7].

There are four procedures, from short to long term, to adapt supply capacity to demand requirements.

1. Extra hours or reduction of hours: This approach is closely related to the "additional shift" process; it allows great flexibility and is used with the advantage of existing employees.
2. Additional shifts: This alternative allows dealing with short-term demand fluctuations. Extra costs and training depend on the demand fluctuation and staff availability.
3. External production: This is used to react to demand peaks; it results mainly in higher unit prices and has the risk of knowledge transfer.
4. Investments: This option requires good predictions of capacity needs, because, as capacity increases, fixed costs also increase.

There are multiple customer implications of a lack of capacity, from delays to market share losses. The consequences of bottlenecks are as follows [8]:

- Limited availability;
- Long delivery times with decreasing reliability;
- Constantly changing priorities and quantities;
- Errors and cancellations;
- High personal commitment of the employees involved in order processing;
- Additional expenditure due to extensive internal clarifications or the search for external alternative suppliers;
- Additional costs through partial and express deliveries, special shifts, and temporary employees;
- Individual decisions instead of overall optimum.

If this condition remains over time, customer satisfaction decreases, and companies lose market share and sales potential. Investments are not immediate and, therefore, do not alleviate the bottleneck in the short term and have the risk of being oriented too much to short-term needs and an overcapacity situation developing in the long term [8].

As a consequence, key indicators for capacity-related decision-making should be defined taking into account that the final goal of each business activity is to increase the value of the company [9]. The main indicators for an investment are return on investment (ROI), the payback period, and the net present value (NVP). Moreover, it is also important to consider the concept of opportunity cost of not making a decision on an investment, for instance, cars not sold due to a bottleneck in the paint shop.

In conclusion, after having calculated those indicators with an assumed demand pattern, other risks should be considered, such as demand volatility, new market entrance, price sensitivity, promotions, etc.

3. Methodology

Cooperation between research partners and the chosen methodological approach was set as a combination of an in-depth literature review, conceptual development, and simulation techniques in a specific case study. The study refers to manufacturing companies with a production network, plant, or group of machines with a certain capacity to serve a specific market.

The method used to reach the goal of matching market demand started from analyzing the structure to develop the conceptual model. For this purpose, the viable system model (VSM) was selected. The VSM is a cybernetic model that was developed by Beer throughout his life [10]. The viable system model is a reference model that can be applied to describe, diagnose, and design management models in organizations [11]. Beer deduced the VSM from cybernetics, and from the central nervous system of the human being, in order to deal with complex systems [12]. Through the analysis of the central nervous system, the minimum requirements that a system must meet to ensure that it is viable were derived [13]. In this way, each sub-system has one of the nervous system's functions. The nervous system is composed of units that execute actions, such as the spine, muscles, and organs (System 1), the spinal cord (System 2), the brain stem (System 3), the diencephalon (System 4), and the cerebral cortex (System 5) [11]. Analogous to the planning levels of a company, the structure of the VSM can be divided into three levels. Therefore, Systems 1–3 (including System 3*) are assigned to the operational level. System 4, in turn, represents the level of strategic planning and stabilizes the entire system thanks to its interaction with the environment. System 5, finally, represents the level of normative planning [11].

Moreover, in order to determine which policies should be used to control the behavior over time and how these policies should be designed and implemented in order to have a robust response against change [14], system dynamics was chosen. Vensim, a software package enabling system dynamics modeling, was selected for this project. Vensim is a software program that allows for the simulation of highly complex and dynamic systems that involve a lot of integrated decision-making. Vensim (Vensim is a registered trademark of Ventana Systems Inc.) provides a high degree of rigor for writing model equations. It adds features for tracing feedback loops. Vensim also provides very powerful tools for the optimization of multi-parametric simulation results, which allows the analysis to validate results and the model's structure, as well as to determine the most convenient policy options by parameterizing these policies.

The innovative approach relies on a combination of the following, in order to generate a decision-making support concept and model in Vensim for balancing supply and demand over time for a specific production system including economic influences:

- The viable system model as a framework for developing the conceptual model;
- System dynamics for designing and controlling the production system's behavior;
- Statistical process control for demand pattern monitoring;
- Forecasting methods for calculating a demand forecast per period.

Once the methods were set, an in-depth literature review of demand planning, capacity planning, and adaptability, along with key indicators for capacity-related decision-making, was performed. As a result, the basis for the conceptual model development was built. It is an integrated model created for all kinds of manufacturing organizations. Later, the generic conceptual model was extrapolated and proven for the case study of an OEM plant. A simulation was programmed in Vensim using assumptions, and validation with extreme values was performed. Finally, different scenarios and decisions were simulated in order to test the hypothesis previously defined.

4. Design and Simulation of the Conceptual Model for Capacity Planning of an OEM Plant

4.1. Design of a Generic Conceptual Model Development Applying the VSM

The VSM environment is represented by customer demand, technological change, quality standards, price, and capacity of competitors. System 5, the normative level of capacity management, determines the goals of ROI and payback standards. System 4 considers the environment and internal set-up and situation in order to make decisions to match supply capacity with customer demand. Decisions are made depending on the reactivity, i.e., the risks to be taken when making a decision, such as making an investment or adding an additional shift. System 3 determines the overall internal planning using a certain forecasting method as well as methods for demand and capacity planning. Moreover, economic parameters are consolidated here. System 2 represents the coordination of the production lines or shops within the production plant to achieve production goals. Finally, System 1 contains the entities or machine groups within a production line or a workshop. In order to support capacity management activities, these groups are viewed as machine groups that can produce the same products. In order to develop the conceptual model, the following activities were performed:

- Development of an economic target system depending on capacity utilization of a production network, plant, line, or group of machines;
- Capacity management tasks according to planning horizon levels;

- Definitions of recursion levels and operative units;
- Association of tasks to recursion levels and operative units;
- Identification of needed information flows between operative units and recursion levels.

The conceptualization of capacity management based on the viable system model, and the tasks related to the five systems of the viable system model are shown in Figure 1.

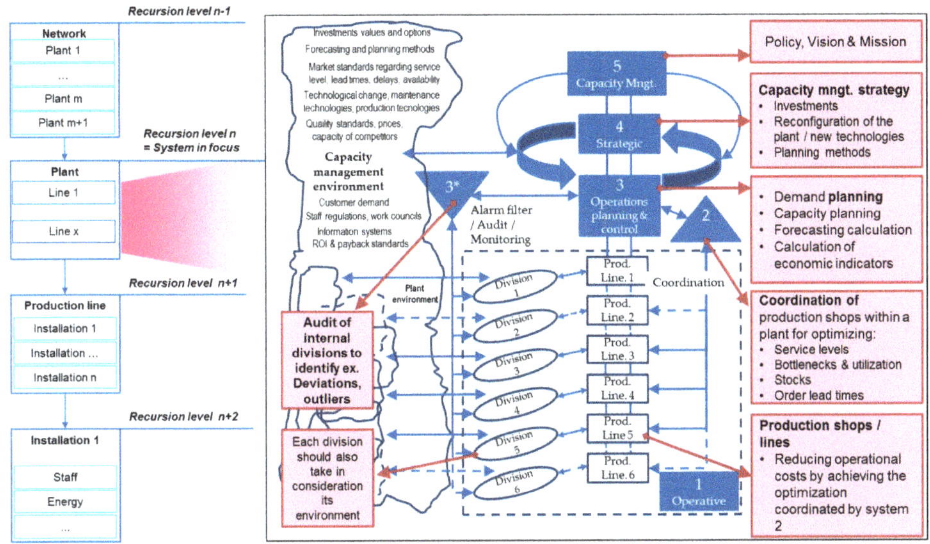

Figure 1. Recursion levels of the conceptual model and capacity management at the plant level based on the viable system model.

In the different recursion levels, casual loop diagrams were constructed in order to identify interrelationships among factors. Moreover, the main characteristics of the conceptual model that applies the viable system model are described below and compared with the classical approach, named the nonviable system model.

Demand planning: There are control limits based on statistical process control principles, such as the time series in quality control charts and hypothesis tests that are used to change a demand forecasting method to another method.

The classical approach uses one method (simple moving average) to forecast customer demand, while the VSM simulation model can change between two forecasting models (simple linear regression and cumulative moving average) depending on the pattern of the demand. Moreover, the VSM simulation model is able to detect outliers, forecast sporadic trends, extend the moving average, and detect seasonal demands.

- The focus of demand forecasting is not to bring about a difference between the models in terms of the accuracy of the used forecast methods, but to be able to respond to the changes in customers' demands by detecting the pattern changes as soon as possible. Due to this premise, the model includes factors that allow for controlling the values that trigger the forecasting changes. Therefore, it is a fight between "detecting only the changes that are real changes" (but perhaps not before they have a negative impact on production and sales parameters) or "detecting more changes than the real ones but knowing that, if there is a change, it is going to be anticipated" (but knowing that, if a change is not a real one, the system will have more forecast failures at first).

- This viable system model deals with uncertainty in demand by calculating forecast values for each product separately. Then, the forecast for groups of customers is done by aggregation of customers' forecasts.

Forecasting models: There are three methods in requirements planning: demand-driven or deterministic, stochastic, and heuristic. Each forecasting method is adequate for a certain demand pattern, as shown in Figure 2 [15]. As mentioned above, one important characteristic of the viable system model is its ability to adapt its demand forecasting. As a consequence, it is able to detect demand pattern changes after it starts forecasting with another method or with the same method but with different parameters. The approach tries to respond as quickly as possible to demand transformations. In so doing, the model can detect fake changes, which would be worthwhile due to its fast reaction capability.

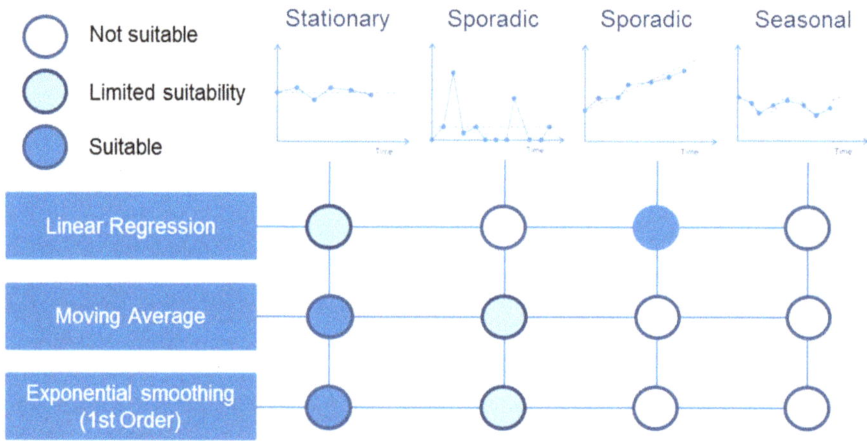

Figure 2. Use suitability of forecasting methods [15].

The VSM model recognizes the following patterns, and it forecasts them with the forecasting methods listed:

- Trend: with regression analysis.
- Trend plus sporadic: with regression analysis plus an average sporadic quantity per day.
- Sporadic: the sum of outliers is divided by their frequency to obtain the average demand per day.
- Steady: with the cumulative moving average method.
- Steady plus sporadic: with the cumulative moving average method plus an average sporadic quantity per day.
- Seasonal: with the cumulative moving average method (however, this is not the appropriate forecast method in terms of accuracy for this kind of demand pattern).

The system is able to detect the following changes by analyzing historical demand data:

- From steady to steady with a lower/higher mean;
- From steady to steady with a lower/higher standard deviation;
- From steady to steady plus a positive/negative sporadic pattern;
- From steady to an increasing/decreasing trend;
- From trend to trend plus a positive/negative sporadic pattern;
- From trend to steady;
- From trend to seasonal;

- From a sporadic pattern to another sporadic pattern.

All of these changes are controlled by the variable called "demand reactivity". This variable affects all of the control limits for the changes described above. When the value of this variable is higher, then the VSM simulation model detects more changes in demand with less accuracy but reacts before real changes occur in the demand pattern. Each of these changes has a lower and an upper control limit. These limits are fixed using a statistical approach that assumes the forecasting parameters have a normal distribution.

It can be seen as a hypothesis test. We use the first type of detection (from steady to steady with a lower/higher mean) as an illustrative example. If the mean of the last 10 days (the sample) for the customer is lower/higher than the mean that is used for forecasting (the population) minus/plus one standard deviation, then a new steady pattern is created with a lower/higher mean, because, with a high percentage, the last 10 values are not a sample of the population (the last demand pattern).

- H_0 = the mean is the same as before: the null hypothesis is accepted when mean of the last 10 values is within the control limits that are the mean plus/minus one standard deviation.
- H_1 = the mean is higher/lower: one of the alternative hypotheses is accepted if the mean of the last 10 values is not within the control limits.

As was explained in the first detection case, the other changes are tracked using the same methodology.

Capacity planning: This has an influence on the decision on the number of shifts, the number of employees, and new investments, each of which has a certain delay until its implementation. The conceptual model compares, for each period of time, the gap between demand and available capacity. Based on it and on the demand planning, which is based on a forecasting method and a statistical analysis, the following measures are considered in the conceptual model:

- Adjust the number of employees in order to produce more in the same number of shifts.
- Adjust the number of shifts, as well as working hours, on weekends or holidays.
- Increase external production capacities for peak or constant production requirements.
- Increase the production capacity with new investments.
- Calculate the daily production loss according to bottlenecks and extrapolate it for longer periods based on the available demand forecast.
- Increase maintenance planning and coordination, as well as the number of tools, in order to reduce breakdowns and increase plant availability.

Economic parameters: Within the conceptual model, three decisions are considered.

- New investments for increasing production capacity. For this decision, the model requires a payback of less than two years. The return on investment is forecast to determine the increase in capacity that is pursued by the investment.
- Operational costs related to extra hours or extra shifts of operational employees. Based on the gap between production capacity and demand, operational costs are optimized by adjusting the number of employees, working days, and shifts.
- Customer demand loss based on the value of the customer order lead time. Customers buy from a company or not depending on the lead time. Therefore, the model assesses the customer requirements and evaluates the losses in sales due to a long delivery process. Based on this loss, a decision on whether to improve capacity and internal processes can be made.

Decision-making: Decision trees are used as a function of the level, i.e., network, plant, line, machine group. For decision-making, a simulation inside the simulation is conducted in order to make a forecast of the decision, taking into account the risk of the decision with a qualitative approach of price sensitivity and new market entrance. To predict the behavior, quality standards for time series are used to analyze the probability that a demand pattern will continue over time (assumption: a demand pattern is only influenced by customer needs, not market influence, normative standards, price, promotions, disasters, etc.).

4.2. Simulation of a Specific Model for the Case Study of an Automotive Plant

The following assumptions were made:

- Time restrictions: Firstly, the modeler must define a time horizon and units of time. It is easy to carry out this step by asking to what extent the simulation should be considered. In the case of the study, it was decided to simulate four working years to evaluate influences in the medium and long term. The simulation was performed with 1000 time periods, each representing one producing day counting in total for four years of production.
- Production capacity has a maximum of 600 units per day in three shifts at the beginning of the simulation. During the simulation, two investment options were considered: an increase of 100 units per day or an increase of 200 units per day. Both models assessed the same requirements for initiating one or the other investment. The difference between the viable system model and the nonviable system model was the lead time of the decision-making process.

  - Viable system model: the decision-making process takes 10 days.
  - Nonviable system model: the decision-making process takes 100 days.

- Demand characteristics and forecast: The model makes different forecasts using two methods, i.e., moving average and linear regression:

  Moving average: the simple moving average (SMA) is the non-weighted mean of the previous n data [16]. In the nonviable system model, it is calculated as the average for the last 10 days ($n = 10$). As the VSM is able to detect demand changes over time, all demand values were taken into account for the forecast until a new demand pattern was registered. Therefore, the formula of the cumulative moving average was applied.

  Linear regression: The regression analysis method is usually applicable to a steady course of a time series or a stable trend [16]. In the model, it is only used for trend demand patterns, when using only one leading indicator and the time since the trend demand pattern was detected. As a result, a simple linear regression can be defined as follows:

$$F_j(t) = \text{demand forecast for customer } j \text{ at time } t = \alpha \times t_{trend} + \beta,$$

  where α is the slope, t_{trend} is the time since the trend demand pattern was detected, and β is the value at the time when the trend demand pattern was detected.

- The existing car model is in a mature stage with stable demand and provides 1000 euros/car of margin. The new model is in the process of being launched and provides 2000 euros/car. These values were used to calculate profits. If there is loss in volume, it is assumed that the new model will have the loss in volume due to unknown future demand.
- The simulation model considers sales loss starting from a customer order lead time greater than 60 days.
- A product is a finished product after it leaves the production facility.
- The warehouses have no stock limitation.
- There is no transport limitation (or limitation to the number of trucks) between the different stages.
- A steady supply of materials for the production process is provided.
- The available stock, as well as the number of products, is known at every moment of the transport process.
- Order information along the supply chain is available.

- Data on historical demand are available for both models one day after the demand.
- Operational adjustments allow for changing the shift model with a one-third increase in the capacity per extra shift with a maximum of three shifts. Moreover, a higher number of employees in an existing shift provide a flexibility of 10% to existing production capacities.
- Production is 50% make-to-stock and 50% make-to-order.
- Production consists of a process that begins with steel stamping and ends with final revision. The plants in the process are shown in Figure 3 and listed below.
- Press shop 1: a nominal capacity of 300 units per day.
- Bodywork shop 1: a nominal capacity of 300 units per day.
- Press shop 2: a nominal capacity of 300 units per day.
- Bodywork shop 2: a nominal capacity of 300 units per day.
- Paint shop: a nominal capacity of 600 units per day.
- Pre-assembly shop, assembly 1: a nominal capacity of 600 units per day.
- Mechanical assembly shop, assembly 2: a nominal capacity of 600 units per day.
- Final assembly shop, assembly 3: a nominal capacity of 600 units per day.
- Final inspection shop: a nominal capacity of 600 units per day.

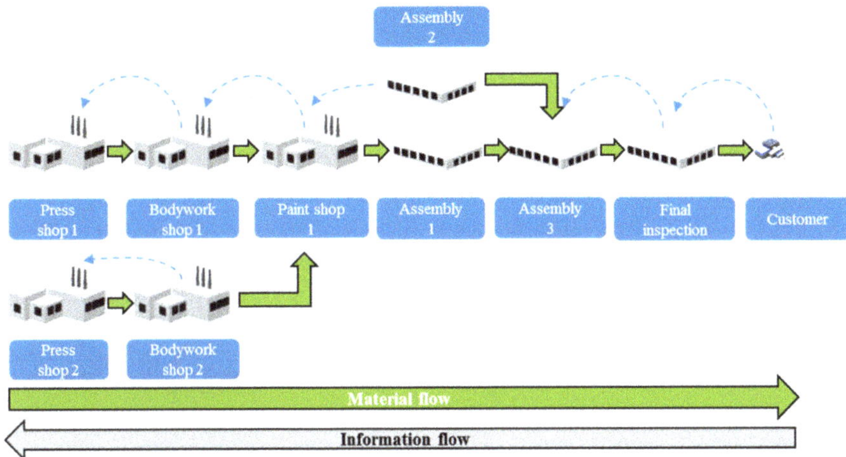

Figure 3. Simulation production flow (own elaboration).

As can be seen in Figure 3, the two car models use the same production capacity for the painting and assembly processes but use different capacities for the bodywork and press processes.

Four demand scenarios were simulated. In all scenarios, there are two products or car models with the same platform. The first model is at a mature stage in its life cycle, and the second model is newly launched with demand depicted in Figure 4. Demand was created in Excel in order to use replication to have exactly the same demand in all of the models. This method of data generation allowed us to create customized demand patterns to be read by Vensim. According to many authors, the basic demand patterns are steady, seasonal, trend, and sporadic demand patterns [16]. These demand patterns were applied in combination as a basis for demand in the models to describe the behavior of the simulation models.

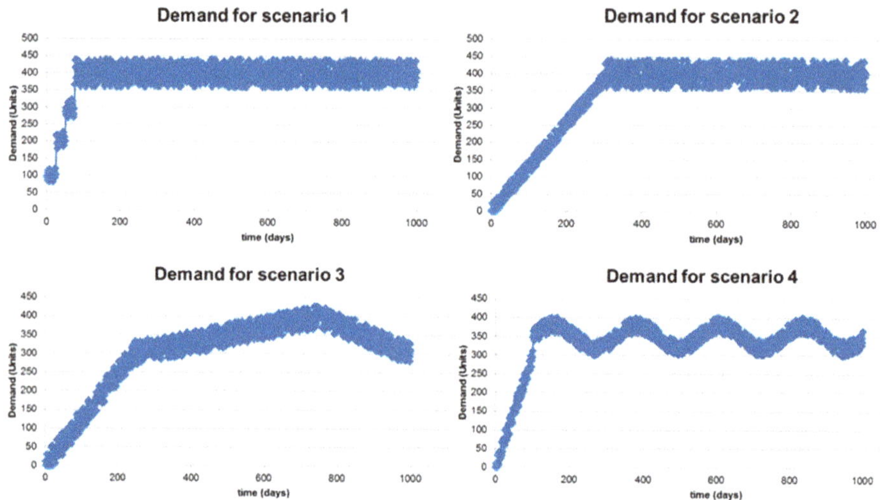

Figure 4. The four demand scenarios for the new car model (own elaboration).

For the four previously described scenarios and both simulation models, the following key performance indicators (KPIs) were calculated from the simulation:

- Profits (million euros): the result of the multiplication of the number of produced cars by the margin that was provided for the type of produced car.
- Total production (units): the cumulative sum of all car units produced over the 1000 simulated production days.
- Capacity utilization (%): the cumulative utilization of available capacities over the 1000 simulated production days.
- Maximum production capacity (units/day): both models were initialized with a demand of 600 units per day. This KPI shows the maximum capacity that was achieved during the simulated period.
- Service level (%): the quantity of units delivered on time divided by the total number of delivered units.
- Customer order lead time (days): the number of days between the placement of the order and the delivery of the product.
- Operational savings (million euros): the savings due to optimization of working hours, shifts, and maintenance activities.
- Investment value (million euros): the amount of the investment made to increase capacities. The value can be 30 million euros for an increase of up to 700 units per day or 60 million euros for an increase of up to 800 units per day.
- Return on investment (million euros): the margin of the products that can be produced thanks to the investment minus the investment value.

4.3. Simulation Results for the Case Study of an Automotive Plant

Before the results from the model were extracted and interpreted, a formal validation was performed. According to Sterman, there are 12 possible methods for validating system dynamics models [17]. There are three that are relevant to our models; one of them is the extreme-value test, which we used in this paper. An application of this method proved that the model's response is plausible when taking extreme values for different input parameters. Some basic physical laws should be examined, for example, when there are no employees, the model's output should not able to meet the demand. For both models, the same input and output variables were chosen in order to analyze and validate the models. These input variables are nominal capacity and customer demand. From the variation in these variables, the following can be expected in order for the results to be logical and the model to be validated:

- For a lower nominal capacity (units per day), the customer order lead time, volume loss, and capacity utilization must be higher, and the total number of units delivered to customers must be lower, as shown in Figure 5. The red lines indicate the lower nominal capacity, and the blue ones indicate the higher nominal capacity.
- For a lower customer demand (units per day), the customer order lead time, volume loss, capacity utilization, and total number of units delivered to customers must be lower.

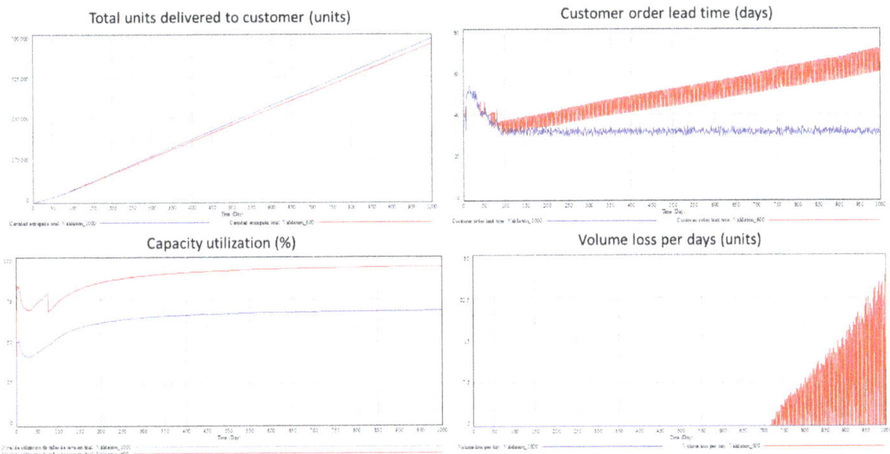

Figure 5. Results of the validation for nominal capacity using the extreme-value test (own elaboration).

After the validation, the results for the four demand scenarios were extracted for the selected key performance indicators.

In the first scenario, the VSM produced better results for all relevant indicators as it can be seen in Table 1, although the nonviable system model increased its capacity by 200 units per day. As this decision was made with a delay in comparison to the viable system model, the total number of delivered units was almost 5000 units lower and the capacity utilization was 12% lower.

Moreover, the viable system model generated a profit of 1021 million euros, whereas the nonviable system model generated a profit of 1012 million euros, which is nine million euros less. In addition, the VSM was able to generate savings of one million euros.

For both scenarios, the return on investment (ROI) was positive at the moment the decision was made. The value was the same; however, the VSM achieved this ROI with an investment value that was 30 million euros less due to a lower increase in capacities than the nonviable system model. As the decision was made later, the customer order lead time was almost 10 days longer in the nonviable system model.

Therefore, from the comparison of the models, the results show that the VSM would be selected as desirable in this demand scenario due to its combination of a low investment and a low cost and its highly efficient deployment (lead time, service level, and capacity utilization).

Table 1. Simulation results for demand scenario 1, stationary scenario, at the plant level. VSM, viable system model.

Simulation Level	Demand Scenario 1	Key performance Indicator (KPI)	VSM Simulation Model	Nonviable Simulation Model
Plant level	- Launch curve until t = 100 days - Stationary demand for the new model at 400 units/day and for the 2nd model at 300 units/day	Profits (million euros)	1021	1012
		Total production (units)	660,129	655,761
		Capacity utilization (%)	94.3	82.0
		Maximum production capacity (units/day)	700	800
		Service level (%)	98.5	97.6
		Customer order lead time (days)	49.7	58.6
		Operational savings (million euros)	1.0	0.0
		Investment value (million euros)	30.0	60.0
		Return on investment (million euros)	70.0	70.0

In the second scenario, the VSM also produced better results for all relevant indicators as shown in Table 2, although, as in the first scenario, the nonviable system model increased its capacity by 200 units per day. As this decision was made with a delay in comparison to the viable system model, the total number of delivered units was nearly 5000 units lower, and the capacity utilization was almost 12% lower.

Table 2. Simulation results for demand scenario 2, trend/stationary scenario, at the plant level. VSM, viable system model.

Simulation Level	Demand Scenario 2	Key Performance Indicator (KPI)	VSM Simulation Model	Nonviable Simulation Model
Plant level	- Trend until t = 250 days until stationary demand for the new model at 400 units/day and for the 2nd model at 300 units/day	Profits (million euros)	934	925
		Total production (units)	617,021	612,520
		Capacity utilization (%)	88.2	76.7
		Maximum production capacity (units/day)	700	800
		Service level (%)	97.2	95.3
		Customer order lead time (days)	46.3	54.4
		Operational savings (million euros)	3.5	0.0
		Investment value (million euros)	30.0	60.0
		Return on investment (million euros)	28.8	15.4

Moreover, the viable system model generated a profit of 934 million euros, whereas the nonviable system model generated a profit of 925 million euros, which is nine million euros less. In addition, the VSM was able to generate savings of 3.5 million euros.

For both scenarios, the return on investment was positive at the moment the decision was made. The ROI in the VSM was 13.4 million euros higher than that in the nonviable system model and it was achieved with an investment that was 30 million euros less. Moreover, the customer order lead time was 8.1 days larger in the nonviable system model due to a long decision-making process that led to an investment to increase production by 200 units per day. The VSM model increased production by only 100 units per day.

Therefore, as in the first scenario, from the comparison of the models, the results show that the VSM simulation model would be selected as desirable in this demand scenario due to its combination of a low investment, operational savings, and highly efficient deployment (lead time, service level, and capacity utilization).

In the third scenario, according to Table 3, the VSM produced better results for all relevant indicators except capacity utilization due to the fact that the nonviable system model did not increase its capacity. As this decision was never made, the viable system model produced a higher total number of delivered units but lower capacity utilization. This situation arises from the fact that, as the demand was volatile and had a trend pattern, the nonviable system model was not able to identify the need for a new investment. This led to a volume loss of almost 45,000 units during the four-year period and a profit loss of 91 million euros. In addition, the VSM was able to generate savings of 10.0 million euros.

As the nonviable system model did not increase its capacity, the capacity utilization was 6.4% higher than in the VSM, which increased its capacity by 100 units per day. The return on investment of this increase is positive and accounts for 70 million euros. Moreover, the customer order lead time in the nonviable system model was almost 30 days longer than in the VSM model. In addition, the service level was also 2.3% higher in the VSM model.

Therefore, from the comparison of the models, the results show that the VSM would be selected as desirable in this demand scenario due to the investment made, the savings achieved, and its highly efficient deployment (lead time, service level, and capacity utilization), thanks to an anticipated investment that the nonviable model did not decide to make.

Table 3. Simulation results for demand scenario 3, trend scenario, at the plant level. VSM, viable system model.

Simulation Level	Demand Scenario 3	Key Performance Indicator (KPI)	VSM Simulation Model	Nonviable Simulation Model
		Profits (million euros)	887	794
	- Trend with slope +1.2 until t = 250, +0.2 until t = 750 and −0.4 until t = 1000	Total production (units)	593,280	546,885
		Capacity utilization (%)	84.8	91.2
		Maximum production capacity (units/day)	700	600
Plant level		Service level (%)	96.3	94.0
	- Stationary demand for the 2nd model at 300 units/day	Customer order lead time (days)	41.0	70.9
		Operational savings (million euros)	10.0	0.0
		Investment value (million euros)	30.0	0.0
		Return on investment (million euros)	70.0	-

In the last scenario, as shown in Table 4, the VSM produced better results for all relevant indicators except capacity utilization due to the fact that the nonviable system model did not increase its capacity. As this decision was never made, the viable system model produced a higher total number of delivered units but lower capacity utilization. This situation arises from the fact that, as the demand was volatile and had a seasonal pattern, the nonviable system model was not able to identify the need for a new investment. This led to a volume loss of more than 51,000 units during the four-year period and a profit loss of 93 million euros. In addition, the VSM was able to generate savings of 12.5 million euros.

As the nonviable system model did not increase its capacity, the capacity utilization was 6.1% higher than in the VSM, which increased its capacity by 100 units per day. The return on investment of this increase was positive and accounted for 20.4 million euros. Moreover, the customer order lead time in the nonviable system model was 40 days longer than in the VSM model. In addition, the service level was also 6.1% higher in the VSM model.

Therefore, from the comparison of the models, the results show that the VSM would be selected as desirable in this demand scenario due to the investment made, the savings achieved, and its highly efficient deployment (lead time, service level, and capacity utilization), thanks to an anticipated investment that the nonviable model did not decide to make.

Table 4. Simulation results for demand scenario 4, seasonal scenario, at the plant level. VSM, viable system model.

Simulation Level	Demand Scenario 4	Key Performance Indicator (KPI)	VSM Simulation Model	Nonviable Simulation Model
Plant level	- Launch curve until t = 100 days - Seasonal demand for the new model at 350 units/day, and stationary for the 2nd model at 300 units/day	Profits (million euros)	935	842
		Total production (units)	617.390	565.968
		Capacity utilization (%)	88.2	94.3
		Maximum production capacity (units/day)	700	600
		Service level (%)	97.9	91.8
		Customer order lead time (days)	44.2	88.2
		Operational savings (million euros)	12.5	0.0
		Investment value (million euros)	30.0	0.0
		Return on investment (million euros)	20.4	-

5. Conclusions

As a result of the research work, the main hypothesis was supported based on the following facts:

- Thanks to a new conceptual model for capacity planning able to make decisions for reduction or increase of production capacity, the viability of a company can be assured. As a result, it proves the need for such a system as a standard tool for managers in the future in order to increase the efficiency and adaptability of manufacturing organizations.
- The viable system model provides the necessary structure to determine the interrelationships among areas and parameters that allow decisions to be made regarding capacity in an autonomous process based on selected control limits.
- The simulation of an OEM plant using the developed conceptual model presents better results compared with currently available structures regarding how to deal with customer demand volatility.

The final goal is to transfer this research method to real production systems, where it could be applied in particular cases as a tool to assist managers by centralizing all data related to a topic in a short period of time, enabling the simulation of what-if scenarios, and speeding up decision-making processes as a unique selling proposition (USP) of manufacturing organizations.

The results show the benefits of capacity management based on a digital-twin approach in which top management can make decisions in a short period of time. However, the question of how it can be made possible in large organizations involving multiple plants and departments remains to be answered.

Future research could develop a virtual model of a manufacturing organization by including other decision-making factors and determine how these decisions and approaches could be realized in the real world with current technologies and systems.

Author Contributions: Conceptualization, S.G.-G.; data curation, S.G.-G.; formal analysis, S.G.-G.; Investigation, S.G.-G.; methodology, S.G.-G. and J.R.; resources, S.G.-G.; software, S.G.-G.; supervision, M.G.-G. and J.R.; validation, S.G.-G.; writing—original draft, S.G.-G.

Funding: This research received no external funding.

Conflicts of Interest: The authors declare no conflicts of interest.

References

1. Kühnapfel, J.B. Vertriebsprognosen. In *Vertriebscontrolling*; Springer Gabler: Wiesbaden, Germany, 2014.
2. Stadtler, H.; Kilger, C. *Supply Chain Management and Advanced Planning*; Springer: Berlin, Germany, 2002–2005; Volume 4, p. 139.
3. Schuh, G.; Stich, V.; Wienholdt, H. *Logistikmanagement*; Springer Vieweg: Berlin, Germany, 2013; p. 89.
4. Campuzano, F.; Mula, J. *Supply Chain Simulation: A System Dynamics Approach for Improving Performance*; Springer Science & Business Media: Berlin, Germany, 2011; p. 23.
5. Schönsleben, P. *Integrales Logistikmanagement: Operations and Supply Chain Management in Umfassenden Wertschöpfungsnetzwerken*; Springer: Berlin, Germany, 2011; p. 668.
6. Frazelle, E. *Supply Chain Strategy: The Logistics of Supply Chain Management*; McGraw Hill: New York, NY, USA, 2002; p. 117.
7. Gupta, M.C.; Boyd, L.H. Theory of constraints: A theory for operations management. *Int. J. Oper. Prod. Manag.* **2008**, *28*, 991–1012. [CrossRef]
8. *Beschaffungsmanagement Revue De L'acheteur*; Verein procure.ch: Aarau, Switzerland, 2010; Volume 10/10, pp. 16–17.
9. Biedermann, H. *Ersatzteilmanagement: Effiziente Ersatzteillogistik Für Industrieunternehmen*; Springer: Berlin, Germany, 2008.
10. Espejo, R.; Harnden, R. *The Viable System Model: Interpretations and Applications of Stafford Beer's VSM*; Wiley: Hoboken, NJ, USA, 1989.
11. Auerbach, T.; Bauhoff, F.; Beckers, M.; Behnen, D.; Brecher, C.; Brosze, T.; Esser, M. Selbstoptimierende produktionssysteme. In *Integrative Produktionstechnik Für Hochlohnländer*; Springer: Berlin, Germany, 2011; pp. 747–1057.
12. Schuh, G.; Stich, V.; Brosze, T.; Fuchs, S.; Pulz, C.; Quick, J.; Bauhoff, F. High resolution supply chain management: Optimized processes based on self-optimizing control loops and real time data. *Prod. Eng.* **2011**, *5*, 433–442. [CrossRef]
13. Beer, S. *Brain of the Firm: A Development in Management Cybernetics Herder and Herder*; Verlag Herder: Freiburg im Breisgau, Germany, 1972.
14. Coyle, R.G. *System Dynamics Modelling: A Practical Approach*; Chapman & Hall: London, UK, 2008.
15. Schuh, G.; Stich, V. *Logistikmanagement. 2., vollständig neu bearbeitete und erw. Auflage*; 2013.
16. Meyer, J.C.; Sander, U.; Wetzchewald, P. *Bestände Senken, Lieferservice Steigern-Ansatzpunkt Bestandsmanagement*; FIR: Aachen, Germany, 2019.
17. Wensing, T. *Periodic Review Inventory Systems*; Springer: Berlin, Germany, 2011; Volume 651.

Article

Value Stream Analysis in Military Logistics: The Improvement in Order Processing Procedure

Raquel Acero [1,*], Marta Torralba [2], Roberto Pérez-Moya [3] and José Antonio Pozo [3]

[1] Department of Design and Manufacturing Engineering, Universidad de Zaragoza, María de Luna 3, 50018 Zaragoza, Spain
[2] Centro Universitario de la Defensa, Academia General Militar, Ctra. de Huesca s/n, 50090 Zaragoza, Spain
[3] Mando de Apoyo Logístico del Ejército, Ejército de Tierra, 28004 Madrid, Spain
* Correspondence: racero@unizar.es

Received: 11 November 2019; Accepted: 19 December 2019; Published: 21 December 2019

Abstract: Military logistics is a complex process where response times, demand uncertainty, wide variety of material references, and cost-effectiveness are decisive for combat capability. The demanding flexibility can only be achieved by improving supply chain management (SCM) to minimize lead times. To cope with these requirements, lean thinking can be extended to military organizations. This research justifies and proposes the use of lean methodologies to improve logistics processes with the case study of a military unit. In particular, the article presents the results obtained using value stream mapping (VSM) and value stream design (VSD) tools to improve the order processing lead time of spare items. The procedure starts with an order generation from a military unit that requests the material and ends before transportation to the final destination. The whole project was structured, considering the define–measure–analyze–improve–control (DMAIC) problem-solving methodology. The results show that the future state map might increase added-value activities from 44% to 70%. After implementation, it was demonstrated that the methodology applied reduced the lead-time average and deviation up to 69.6% and 61.9%, respectively.

Keywords: military logistics; lean management; DMAIC; VSM-VSD

1. Introduction

Logistics and supply chain management (SCM) propose a challenging issue for worldwide organizations. In particular, military logistics [1] is a specific case where the ability to provide human and material resources is crucial in terms of minimum time, unpredictable quantity, and variable location of new armed conflicts. According to [2], five main aspects are different between military and consumer supply: large numbers of different types of items, variable demand, supply management considering priority matters (e.g., medical supplies, subsistence, repair part), the necessity of equipment and supply readiness, and different theatres characterized by moving points. Hence, it is difficult to manage inventories under these market conditions, which could mean either overstocking or delay. Additionally, two more factors must be considered in military applications. The first one is related to the available funding that may be limited for each nation. The second one is the supply importance if it supposes a risk factor to the human lives of combatants and civilians. In other words of the American report [3], the strategic approach of military logistics should be focused on the improvement in processes, information systems, organizational structures, and advances in distribution and transportation technologies. Thus, precise time, capacity, and efficiency of delivery to the operations theatres are required for Armed Forces. As it is explained in [4], in 2001, the United States Department of Defense (DoD) began the standardization process of its SCM using the supply chain operations reference (SCOR) model [5,6] as its framework. This approach focused on the increase in the supply chain reliability by synchronizing each element internally and externally, forecasting the demand and managing inventories and assets efficiently.

To minimize response times and to assure the required flexibility avoided wastes, the lean philosophy can be extended to supply chain management in military organizations. For example, the United States Department of Defense (DoD) has been working intensively together with service providers and supporting contractors to introduce and apply lean principles into their organizations to optimize internal lead times [7]. The origin of lean philosophy is generally attributed to the practices developed from the TPS or Toyota Production System [8,9], pioneered by Taiichi Ohno [10] and Shigeo Shingo [11]. Their lean principles related to philosophy, processes, people, partners, and problem-solving allow organizations the implementation of lean thinking at different levels. Moreover, TPS has influenced not only manufacturing concepts but also supply chain management ones [12–14]. The concept of lean supply is described in [15,16] as an operating attitude that needs to be changed in relation to suppliers so that the effect of associated costs to non-perfect processes will not be limited to the location of the execution. This approach targets long-term customer satisfaction. Thus, supply chain optimization is possible according to the three main TPS goals: best quality, lowest cost, and shortest lead-time, which are achieved by continuous improvement and increasing operations' added value.

This work evaluates the use of lean methodologies and their application to military logistics functions, focusing on supply chain management processes for spare parts. We present a case study to improve military material order processing procedures by implementing value stream analysis methodologies. SCOR model performance metrics, such as order fulfillment lead time and delivery performance, are assessed. To structure the research project, the define–measure–analyze–improve–control (DMAIC) Six Sigma methodology was followed [17]. Its aim is aligned to continuous improvement and lean thinking [18] and integrates the approaches of lean and Six Sigma as presented in [19–21].

2. Background

2.1. Military Logistics and Processes

The supply chain management concept is a horizontal strategic function that encompasses all the operations of the supply chain between customers and suppliers, distribution, manufacturing, procurement, and planning, to give an integrated answer to the competitive difference of the organizations. Viewed from the life cycle perspective, military logistics is the bridge between the deployed forces and the industrial base that produces the weapons and materials that the forces need to accomplish their mission [22]. The North Atlantic Treaty Organization or NATO defines logistics in [23] as the science of planning, coordinating, and carrying out the movement and maintenance of forces, which covers aspects of military operations related to material, transport of personnel, acquisition of facilities and services, and medical service support.

Three main aspects of logistics need to be highlighted in the military logistics cycle: production or acquisition logistics focused on the procurement of the material; in-service logistics that links production and consumer logistics and comprises functions associated with receiving, storing, distributing, and disposing material to the force; consumer or operational logistics that concerns the reception, storage, transport, maintenance, and disposal of the material, including the provision of support and services. Depending on the strategic, operational, or tactical level of the logistic function approached, different activities are encompassed. There are also multiple logistic functions identified in the military field. Nevertheless, we will focus on the material supply logistic function in this work, which includes the determination of stock levels, provisioning, distribution, and replenishment restricted in this case to operational and tactical levels.

The organization of this research study is referred to as ABC throughout this paper due to non-disclosure requirements. Particularly, ABC is a military unit of the Spanish Army in charge of the supply and maintenance of a wide range of items used as spare parts. Therefore, these supply and maintenance processes could be classified into operational and tactical levels. There are several particularities that make the military supply especially critical [2]. In brief, the large number of different material references, variable demand and multiple ordering locations, priority requirements and high material availability are reasons to justify the necessity of optimizing the material order processing procedure and its metrics in ABC. The process to be optimized starts with an order generation from a deployed military unit that requests the material to ABC, and it ends before the transportation to the final destination [24]. Additionally, for the ABC product families, a critical aspect need to be highlighted, i.e., the high rate of technological obsolescence of the material.

The importance of spare parts management in organizations has been addressed in the literature. Supply chain management for spare parts is described in [25] as a multi-echelon supply chain, where the differences in size generate demand peaks and, thus, a very variable and lumpy demand pattern. The authors propose an algorithmic solution considering the sources of demand variability, a probabilistic forecast, and inventory management. The importance of spare parts logistics is mostly related to their inventory management, whose main differences to general inventory management are the low demand and a wide variety of items with no predictable demand. The authors in [26] discuss the basic principles affecting the management of spare parts logistics, which affect the strategic choices and related policies in this area. The most relevant control characteristics of spare parts (criticality, specificity, demand pattern, and value) are first identified to define further supply chain management strategies. The Norwegian Defense has used a systematic approach based on OPUS10 for spare parts management optimization in procurement projects. This case study was reviewed by analyzing empirical data to evaluate the suitability of the theoretical system approach used through OPUS10 in terms of spare parts costs and availability [27]. Furthermore, in [28], the importance of the supply chain as a source of commercial-military integration linking defense production to the wider economy is demonstrated.

2.2. Lean Methodologies and Tools

Lean thinking promotes a continuous-improvement culture with its tools and practices widely applied in different sectors and organizations. Lean is one of the most influential recent paradigms in manufacturing, also considered in relation to another promising trend as Industry 4.0 [29]. It has expanded beyond the original application to other areas and sectors, either public or private. As before mentioned, the origin of lean philosophy generally attributed to the practices developed from the Toyota production system [8,9]. One of the first studies mentioning lean concepts comparing automotive manufacturing plant performance in Japan, the United States, and Europe was [30]. Since then, lean practices have been developed and extended worldwide. The lean approach, its techniques, and limitations were revised in the literature [31,32], remarking the necessity of an adequate implementation sequence [33] and the effect of large-scale strategic management in lean deployment [34]. Several efforts were focused on the measurement of leanness by considering different dimensions [35,36]. The evolution of lean was also addressed in [37] not only as a concept but also in terms of its application.

The link between lean and the supply chain evolved from the value stream concept [38], and the concept of pull was extended beyond single manufacturing facilities to include the up- and downstream partners. Focusing on supply chain management, different methodologies have been applied for logistics optimization purposes. Combined approaches of lean and agile methodologies applied to SCM are also found in [39] for the textile and apparel sector, where short product lifecycles, high volatility, low predictability, and a high level of impulse purchase have paramount importance. In addition, in [40], the existence of hybrid supply chain strategies with a mixed portfolio of products and markets where neither pure agile or lean strategies apply is remarked. Lean works best in high volume, low variety, and predictable environments; meanwhile, agile suits for less predictable environments, with volatile demand and high requirements for variety. An integrated proposal of both methodologies is proposed in [41] and in [42] supported by a personal computer (PC) supply chain case study.

In relation to lean logistics in defense, there is scarce literature available. The concept of pull systems applied to military logistics was questioned by the authors in [43]. A large accumulation of stocks in intermediate distribution points supposes the reduction of effectiveness maneuverability of the deployed combat forces. On the other hand, the "Just-in-Time" approach could bring an associated risk of late or even null deliveries, given the possible actions of the enemy. Bean et al. [44] discussed the inventory management in the uncertain environment of military support. Considering stocks, additional problematic aspects are related to damage, degradation, and obsolescence of the material, which mean monetary losses at the end.

Value Stream Practices

The origin of the value stream analysis, i.e., value stream mapping (VSM) and value stream design (VSD) methods, is the Toyota production system. It is a visual tool and facilitates the continuous improvement in processes efficiency with the identification of value-adding activities and the elimination or modification of adding waste tasks. Early publications showed during the 1990s the benefits of value stream analysis as an operational approach for a lean enterprise and defined specific lean tools for minimizing the seven wastes [45]. Although it is a simple and standardized methodology, it presents limitations and challenges, as was reviewed in [46]. Generally, four steps should be established to improve a process by using VSM and VSD tools. First, a particular process for a product or product family should be selected due to particular factors (e.g., criticality, impact, efficiency, etc.). Then, value stream mapping starts analyzing and drawing the current state map of the process, where activities with added and non-added value are exposed. The different activities, materials, and information flows are related and schematized in flow diagrams after walking along the actual process. The main key performance indicators (KPIs) of the process, such as lead-time, are usually measured before VSM to establish a reference point of the initial situation. Value-added, non-value added activities, and inventories are evaluated in terms of their time contribution to the lead-time of the process. After VSM, value stream design outlines the idealized solution for improving the studied process to reduce waste, lead-time, work in process (WIP), and inventories. At this phase, KPIs are measured again, and an action list of improvement proposals is attained. To implement the updated procedure progressively, a pilot series can be set up to validate the measures applied. Finally, the work plan and final implementation are carried out.

VSM/VSD methodology has been widely applied in multiple organizations and industries. Several articles have been reviewed in the literature, not only considering specific industries [47] but also including different sectors [48,49]. Processes of a wide number of case studies demonstrated the benefits of this lean technique. For example, from the automotive and transportation industry, Wee et al. presented VSM as an effective tool to systematically analyze a lean supply chain problem [50]. Lead time and cycle time were reduced in process lines of auto-parts [51,52] and manufacturing cells [53]. Other manufacturing sectors with confirmed performance results after VSM/VSD applications are plastic injected products used in the healthcare industry [54] and textile and apparel companies [55]. Authors in [56] analyzed traditional steel production processes used in appliance manufacturing and demonstrated the suitability of VSM lean technique and simulation models to evaluate the implemented configurations. Additional case studies with VSM/VSD application with satisfactory results were the industrial paint manufacturing case presented in [57] and the fishing net production company of [58]. Nowadays, a new approach of VSM/VSD related to Industry 4.0 is also under study [59], considering the modeling of internal logistics data [60] or integrating the VSM methodology with a system dynamics analysis [61].

Supply chain challenges in the consumer goods sector were addressed in [62] and [63] to improve two particular processes in the wine and agri-food industry, respectively. The success of value stream mapping is, for example, also demonstrated in product development phases [64], logistical system design [65], software development [66], and service industries [67], so that the universality of this lean technique is worthy of note. Finally, other applications of value stream analysis focused on environmental, waste reduction [68], or manufacturing sustainability performance improvement [69–71] are presented in the bibliography, and also social and economic sustainability issues are targeted with value stream analysis [72–74].

Additionally, new approaches for VSM have been presented in the literature. Haefner et al. [75] included in the VSM analysis quality assurance measures to reduce the rate of defects and quality-related costs. Toivonen et al. [76] introduced in the value stream analysis innovative principles, such as TRIZ and ideation tools, for improving complex processes with a holistic understanding of systems. In [77], a holistic and multi-level VSM approach is presented for multiple sectors. The authors in [78] proposed a similar VSM analysis but, particularly, for information streams in a demanding production case environment.

However, thorough research revealed the scarce academic publications on the evaluation of lean techniques, such as VSM/VSD, applied to the military field and to their logistics processes. In this work, the case study addresses the use of lean tools, such as VSM/VSD, in military logistic processes to improve the material order processing lead-time as a key performance metric of the ABC organization.

3. Methodology

To structure the research, DMAIC (define–measure–analyze–improve–control) methodology was followed, whose principles are aligned to kaizen or continuous improvement in lean thinking. DMAIC is a structured procedure used in Six Sigma and often described as a problem-solving approach [79] to improve manufacturing and business processes by minimizing their variability when focusing on defects and their causes. Six Sigma projects using DMAIC and their benefits are widely demonstrated in the literature [79–83], even with application to SCM [83,84]. Some examples of Lean Six Sigma in military context were found. In [85], the author analyses the application of Lean Six Sigma (LSS) within the Department of Defense (DoD), providing some examples of implementation of this methodology in the United States army, navy, and air force. In the same direction, Baily et al. [86] applied LSS in an army depot maintenance and support processes of command and control systems across the Department of Defense (DoD), reducing the number of repeated material runs for end items and raw materials by 50% in the depot machine shop.

In this work, value stream analysis was integrated into the Six Sigma DMAIC structure for the project, as is shown in Figure 1. We defined a list of the activities planned in the project that was structured into the five DMAIC phases. In addition, a definition of the lean tools selected for the project was done, focusing mainly on Gemba Walk for the process audit, and value stream mapping (VSM)–value stream design (VSD) for the supply process analysis and optimization.

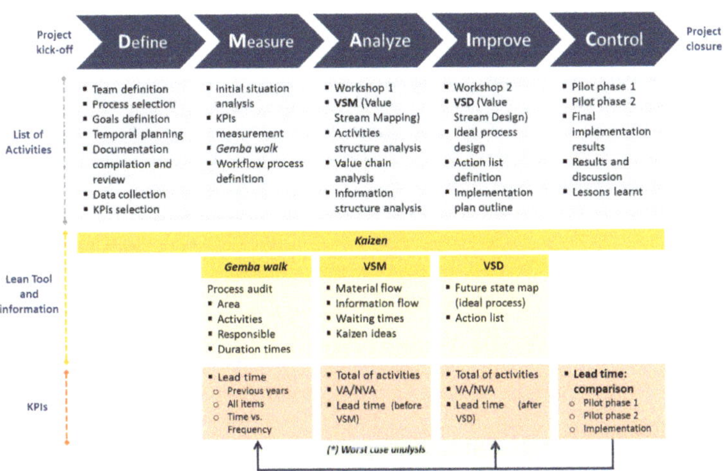

Figure 1. Define–measure–analyze–improve–control (DMAIC) project structure: lean tools and project activities.

4. Case Study

As previously said, ABC is a military unit in charge of the supply of a wide range of spare parts. Due to the importance of response times for armed forces purposes, the selected process to be improved is the order processing of spare parts and new materials that military units deployed in national and international locations demand. Therefore, the study is focused on the procedure between the order generation from a military unit that requests the material and transportation to the final destination.

In the Define phase of DMAIC and after the constitution of the project team, the process under study was selected, and data collection was done from the registered information in the main database of ABC. In this case, we considered the orders, which are processed daily by the ABC unit. The key performance indicator (KPI) of the process to be optimized is the lead-time of the order fulfillment. Thus, it was assumed the lead-time of the process was the sum of not only the order's processing time but also waiting times that could produce delays or intermediate stocks along the process. Due to the different activities carried out in the process, the complete lead-time was also subdivided into several intermediate times that were measured. In particular, the lead-time used in this work was calculated from the date when ABC received the order until the date when the delivery notification was sent to the requesting military unit.

Once the process to be analyzed and its reference KPIs were defined, the goals of the project could be established. The organization ABC initially proposed to reduce the lead-time of the order processing at the end of the project up to 0.5 days. Additionally, the following requirements and targets are settled:

- Non-value added activities in the supply process should be identified.
- The affected areas of the process and the required times of their activities should be evaluated.
- General process improvements should be identified and the effectiveness of the actions proposed should be validated.
- The new procedure should be documented by redefining activities, workflow, responsibilities, and layouts.
- The requirements established by their own organization should be guaranteed.

The analysis of the current performance of ABC was carried out in the measure phase based on the historical data of the years: 2014, 2015, and 2016. The ABC unit handles an average of 15,000 orders per year, which could correspond, independently, to requests from a national territory or from the operation zone. Based on the origin of the order, a priority is assigned. The initial study calculated the average lead-time of the process before the lean optimization to establish a reference value as a zero point for the lead-time indicator. The values calculated for 2015 were 6.76 days for the average lead-time and 6.74 days for the standard deviation of the process in the study. It is worth mentioning the high dispersion of the values measured. Therefore, an additional target for minimizing the high variability of the process was settled.

Additionally, the main incidences in terms of time and frequency, material parts affected, and their potential causes were identified. In this way, an analysis of critical materials was carried out for the most representative period (January to June 2016). It can be concluded that 5% of the material orders were completed on the same day that the supply order was generated from the unit. However, there were critical materials with relatively high values of lead-time. As shown in Figure 2a, among all the materials ordered from January to June 2016, 3% of those material numbers had a lead-time greater than 9 days. This value has been considered as critical after the analysis of lead times in years 2015 and 2016, in terms of average lead-time and standard deviation. These relatively high average lead time values resulted from having materials with high peak lead time values and low ordering frequency, or high ordering frequencies with more moderate lead times. The representation of the different areas cited according to the problems associated with the material (high lead-time or high ordering frequencies) is shown in Figure 2b. These materials were analyzed individually, and specific action plans were defined, given their big influence in the global lead-time indicator of the process.

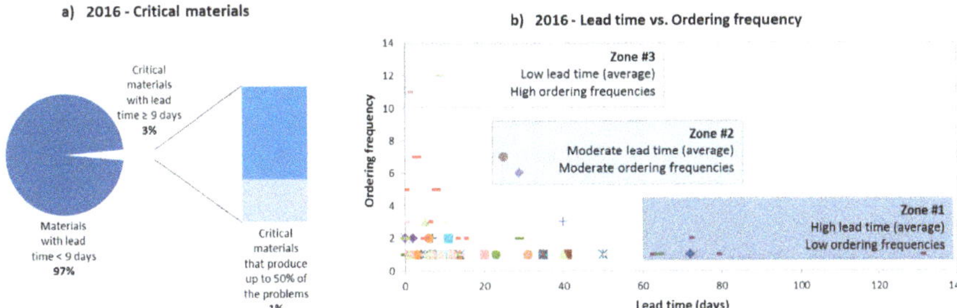

Figure 2. (**a**) Critical materials global analysis (2016); (**b**) Lead time vs. ordering frequency of critical materials (2016).

4.1. Current State Map

Gemba is the lean term related to the "go-and-see" principle that refers to "real place", where the work is happening. Hence, Gemba walk is the tool to analyze processes with concerned people by observation at the value-add location. This also served as an opportunity to discover Kaizen ideas, whereas the different areas, managers, tasks, and estimated times are recorded from the last process activity to the first one. In this case study, three departments of the supply chain were audited: the supply chain control office, the internal material warehouse, and the expedition area to the final military unit. There are approximately 15 employees working in these departments, with 8 am to 3 pm working shifts five days a week. Gemba walk finished with the workflow diagram generation of the material order process in the ABC unit. This diagram represents all the activities revised given the affected departments and considering the critical path. This critical path is the worst-case process in terms of time (longest lead time) and number of tasks (highest number of activities and intermediate waiting times).

Integrated into the analyze phase and after the workflow diagram definition, the activities structure was analyzed. In this case, 48 activities encompass the current material ordering process, being 21 classified as value-added (VA), 21 as non-value added (NVA), and 6 as semi-value added (SVA). In addition, different types of wastes were identified in the process, representing the over-processing of the material order in the different areas, a percentage of 54% out of the total mudas or wastes identified in the process. It was clearly seen that overwork or redundant verifications were handled systematically along the process. Other remarkable wastes were the waiting times between operations or sub-processes (22%) and the reworks (13%) due to failures in the process. Based on this analysis, one of the focuses in the VSM/VSD was to eliminate redundancies that affect clearly the final lead-time of the process.

Once the classification and time of all the tasks were defined, and the complete workflow finished, the process timeline was obtained, and the key performance indicators of VSM were calculated. This is shown in Figure 3, where the ABC current state map is simplified to summarize the results obtained: total and added value activities, lead-time, and kaizen ideas associated with each area of the process.

The VSM confirms the excessive material transport and high administrative workload detected in the process because of the established organization procedures. Nevertheless, both aspects might be improved in the ideal case. In summary, five waiting times were identified in the material order processing procedure of ABC representing these inventories 13.84 days that the material order could be blocked in a worst-case scenario. This could be due to delays in order's endorsement from superiors, pending transport approvals, or further calibration and testing of the order's material. Over the VSM represented in the swim lane diagram, 20 kaizen ideas were tagged. The sequence of the activities presented in the current state map supposes a lead-time of the process of 49.73 days. If the relation between added value (VA) and non- and semi-added value (NVA + SVA) is calculated, the obtained percentage is 1.57%. This result indicates the high improvement margin that exists in the process, taking into account the small percentage of activities effectively devoted to the processing of the order and material requested.

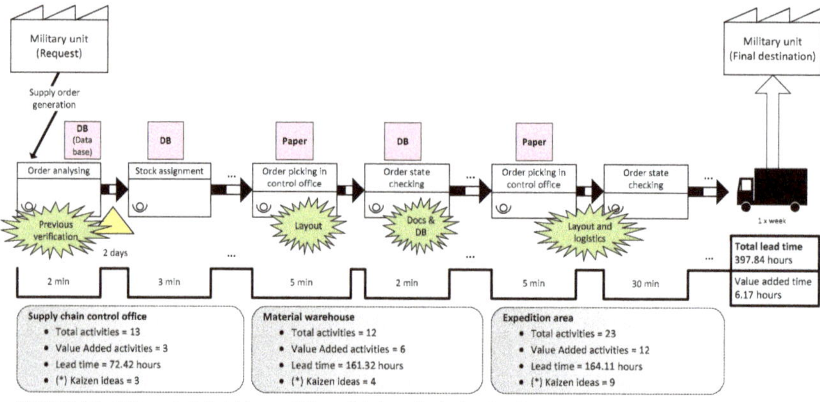

Figure 3. Value stream mapping (VSM) for material order processing procedure: simplified current state map.

The information flow and the mechanisms used are also a key issue to be analyzed and optimized with the value stream mapping. Those mechanisms were indicated in the current state map, and the potential ideas to improve the value-added in the information tasks were included in the action list. The information flow from customers, the military unit that requested the material in this case, to the ABC unit, was standardized in the official information technology (IT) logistic information system of the army. The ABC personnel proposed a vast range of optimizations in the information loading process in the IT system. In addition, hard copies of the IT material order registered in the system were unnecessarily printed, used as working paper along the process, and finally stored. This could be easily solved by working only with the soft order in the IT logistics system. In this line, massive changes in the main SCM information system and future elimination of additional existing databases were considered.

4.2. Future State Map

The design of the ideal process consists of defining and listing the activities that could improve the material order and information processing flow, according to the critical points and potential upgrades identified in the VSM analysis. All these activities are covered in the improve phase of DMAIC. The degree of improvement is measured by the indicators, i.e., lead-time of the process. Hence, to meet the requirement of minimizing the lead-time, we carried out different actions to increase the value of the process. This was achieved by involving people and key partners, optimizing the current activities, and eliminating and/or minimize waiting times and inventories between tasks to create flow [34].

With the new list of process activities, the future state map out of value stream design (VSD) is schematized in Figure 4, where the improvement obtained in the total lead-time of the complete process is worthy of note.

The comparison between the indicators measured before VSM and after VSD is shown in Table 1 and Figure 5. The number of activities was reduced up to 56% from 48 to 27, by minimizing NVA activities (from 21 to 5) and SVA activities (from 6 to 3). Therefore, the value-added of the process was increased. Considering the number of activities, VA tasks suppose 70% of the whole process after VSD. The ratio between added value and non-added value operations is for the ideal situation, 48.86%. This shows the optimization in terms of value in the whole process. The lead-time of the process was also minimized from 49.73 days to 0.75 days. In brief, the presented results indicate that the order processing procedure could be noticeably improved theoretically, by implementing the proposed actions. As a result, the ABC unit would be able to complete an order in less than one day in the ideal process situation.

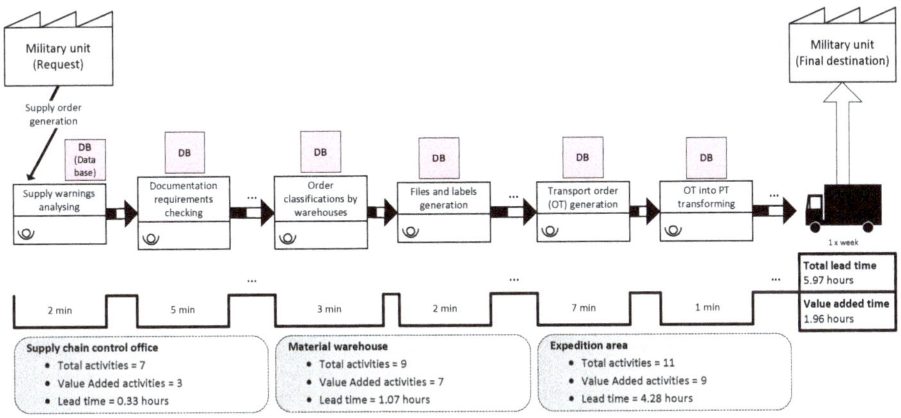

Figure 4. Value stream design (VSD) for order processing procedure: simplified future state map.

Table 1. Process indicators before value stream mapping (VSM) and after value stream design (VSD).

Indicator	Before VSM	After VSD
Total activities	48	27
VA activities (%)	21 (44%)	19 (70%)
NVA activities (%)	21 (44%)	5 (19%)
SVA activities (%)	6 (12%)	3 (11%)
VA (hours)	6.17	1.96
NVA (hours)	391.68	4.01
VA/[NVA + SVA] (%)	1.57%	48.86%
Lead Time (hours)	397.84	5.97
Lead Time (days)	49.73	0.75

Notation: VA (value added), NVA (non-value added), SVA (semi- value added).

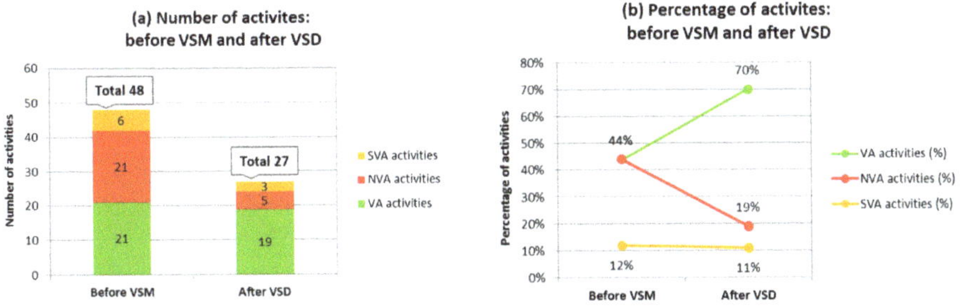

Figure 5. Results before VSM and after VSD: (**a**) Number of activities comparison; (**b**) Indicators of activities percentage.

Both kaizen value-stream improvement focused on material and information flow and process-level kaizen consisting of elimination of waste at the shop floor level were targeted. The summarized outcomes of the future state map are the following: (i) optimize material transport routes and procedures between warehouse and expedition area, (ii) eliminate redundant material checking and identification tasks, (iii) establish new standardized procedure for daily routine between the supply chain control office and the warehouse, (iv) personnel capacity balancing among departments based on tasks redesign and new assignment plan, (v) design of new material pick-up routes in the warehouse based on material location and volumes, (vi) generation of change requests package for routine administrative tasks in IT army logistics information system affecting the material order processing procedure, (vii) improve the information system of the process by elimination of material order printed documentation, (viii) elimination of waiting times between departments affected, (ix) monitoring and control of primary indicators of the process systematically, (x) increase visual management in the whole process.

4.3. Implementation Plan Outline

The kaizen improvement ideas were included in the action list, where responsibilities and affected tasks were defined. Thus, there were 20 actions to be implemented to achieve the ideal process defined in the VSD. All of them were carefully analyzed by the organization and the owner of the process to evaluate their impact, resources needed, and potential risks to be avoided using a contingency plan. Ideas that demanded capital expenditure were limited since the study was carried out in the middle of the year, and the budget of the ABC unit was limited. This analysis ended with the definition of the implementation plan for the new actions.

Value stream design concludes with the theoretical results obtained for the new ideal process. Nevertheless, to validate these results, the implementation of this new, improved process should be attained. As the different proposals for process change were not easy to implement at once and to validate them progressively, two pilot phases of implementation were defined in ABC before the complete deployment of the action plan in the whole process. Both pilot phases were extended in time approximately for 10 working days. The affected number of orders and associate material references were incremented progressively from phase 1 to phase 2, being arbitrarily selected in both phases. In brief:

- Pilot phase 1: 11 working days, 26.4% of the total request orders processed with the new process, 143 different types of material references.
- Pilot phase 2: 10 working days, 53.0% of the total request orders processed with the new process, 200 different types of material references.

Thus, in the next DMAIC control phase, we included the analysis of the indicators obtained in the pilot implementation phase as an important input to confirm if the project copes with the initially established requirements.

At the end of pilot phase 1, the method used for evaluating the effectiveness of the improved process implementation was hypothesis testing or sometimes referred to as significance testing. Hence, the research hypothesis (H_0) was that the new procedure is related to the minimization of the lead-time value. Two groups were defined, given the processed orders considering procedures, the previous one and the new proposed. Additionally, pre-test and post-test differentiation were made with regards to orders included before and during the pilot phase 1, respectively. A parametric test was considered due to the accomplishment of the following assumptions: randomness of the sample observations, normality (central limit theorem), homogeneity of sample variances from the same population. A Student's *t*-test was applied, considering the 95% confidence interval to test a hypothesis about two means. Table 2 shows the results of the test, including the lead-time by orders. Given the significance obtained, in both cases, the value of 0.05 was not exceeded, H_0 could not be rejected. Then, the pilot phases continue with phase 2.

Table 2. Significance testing results considering pilot phase 1.

Study Groups	*t*-Test for Equality of Means	*t*	Freedom Degrees	Sig. (2-Tailed)	Mean Difference	Std. Error Difference
Post-test Experimental group and control group	Equal variances assumed	−3.56	608	0.000	−0.614	0.172
	Equal variances not assumed	−3.50	518.9	0.001	−0.614	0.176
Experimental group Pre-test and Post-test	Equal variances assumed	5.43	10939	0.000	2.243	0.413
	Equal variances not assumed	18.18	733.2	0.000	2.243	0.123

5. Results

The KPIs obtained for both implementation phases are shown in Figure 6, considering the total number of orders processed by ABC during the year 2016, and the ones processed in the two pilot implementation phases. The lead-time was significantly improved in the new VSD process (pilot phase 1 and pilot phase 2). If we compare the current and implemented process (2016 data), the lead time mean value was reduced by half approximately, as can be seen in Figure 6. Additionally, its deviation was also improved, achieving a reduction value of 76%.

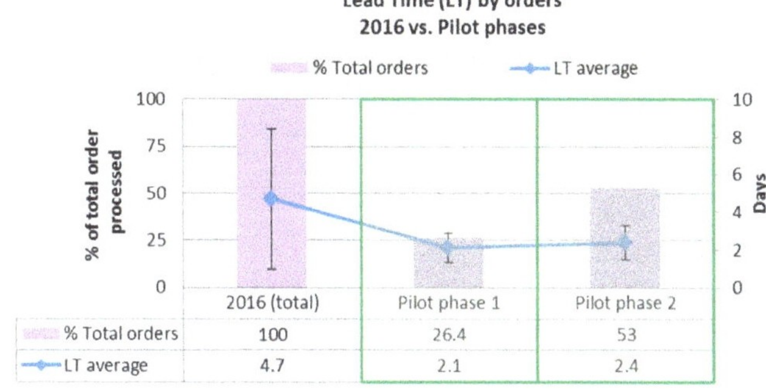

Figure 6. Pilot phases results: lead-time by orders.

Due to the fact that the different orders might not be comparable in terms of requested and type of material, a specific analysis considering material references is also presented. In this case, the material references included in each pilot phase were compared to the same material references behavior with the old process before VSM/VSD (2016 data in Figure 7). As is shown in Figure 7, the improvement in the average lead-time and deviation was proved in both pilot phases. In pilot phase 1 (see Figure 7, left), the results showed an improvement in the average lead-time of 45.24%, and the standard deviation of the lead-time value decreased 39.13%. During the pilot phase 2, when the new process was extended to more material orders processed by ABC (see Figure 7, right), the average lead-time improved 50%, and the dispersion of the process measured with the lead-time standard deviation indicator showed a 29.17% improvement. As it was mentioned before, it was not only important to reduce the lead time of the material ordering process of ABC to optimize their fulfillment delivery to the external units but also to minimize the variability of the process. It is important to highlight that no increase in the orders reject rate was detected after the implementation of the VSD new process in the pilot series. Figure 8 represents the improvement achieved by comparing the lead-time vs. ordering frequency by material reference before and after implementation pilot phases. Despite the limited number of references and quantities analyzed (y-axis), the lead time values calculated for the materials included in the orders processed under the new procedure showed a clear reduction. It is proved that for the same materials the actions considered provided better performance results in terms of delivery fulfillment.

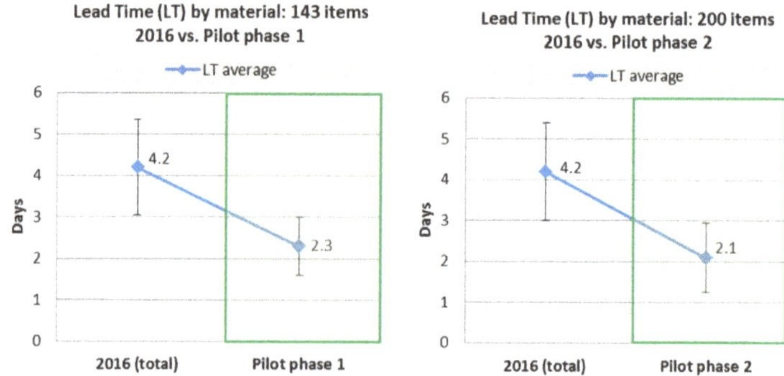

Figure 7. Pilot phases results: lead-time by material reference.

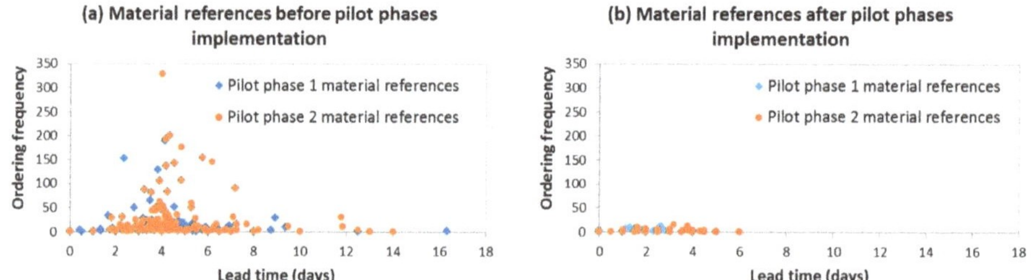

Figure 8. Pilot phases results: Lead-time vs. ordering frequency by material reference (**a**) before and (**b**) after implementation pilot phases.

As a last step of the VSM/VSD, the new procedure designed with the VSD, validated with the two pilot implementation phases, was applied to the 100% of the material orders from January 2017 onwards covering approximately 7800 orders after considering all the action list proposals. The following results confirm the positive degree of improvement in the ABC material order processing procedure with the application of lean methodologies. The temporal evolution in the lead-time (average and standard deviation) is represented in Figure 9, where the relative variation is shown in percentage considering the last four years (2014–2017). It is demonstrated that the KPIs of the process, including the VSM/VSD results, show an improvement. After the implementation phase in 2017, the average lead-time and deviation were reduced up to 69.6% and 61.9%, respectively, considering the initial situation in 2014. The results also confirmed the data obtained in both pilot implementation phases. Therefore, the KPIs behavior in the year 2017 after the complete implementation of the VSM/VSD actions in the process improved successfully.

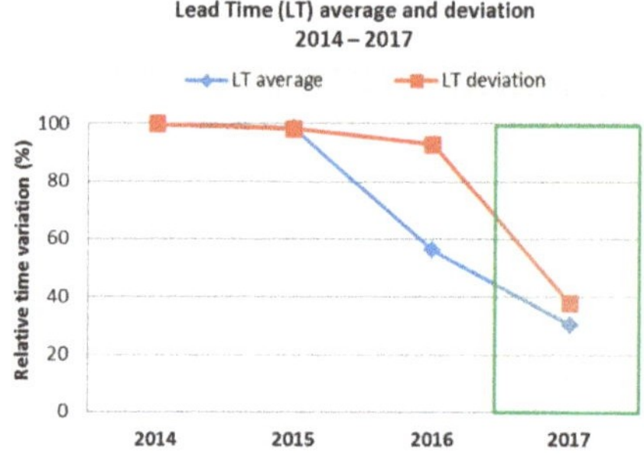

Figure 9. Temporal evolution in percentage terms of average lead-time and deviation (2014–2017).

Thus, we can conclude that the application of the lean tools VSM/VSD to the material order processing procedure carried out in the ABC military organization was correctly implemented, clearly improved the average lead-time of the process, and reduced its standard deviation showing an evident improvement in its variability.

6. Discussion

This work presents the implementation of lean methodology value stream analysis in a particular case of military logistic processes. Specifically, the lean methodology value stream analysis (VSM/VSD) was applied satisfactorily to the material order processing procedure, which plays a key role in the ABC military organization supply chain. First, the state-of-art study showed that there is a lack of references with the same purpose and field of application. Although lean tools, such as VSM/VSD, are widely used in the industry, military issues are not so common in the literature when relating lean thinking and logistics. In addition, the case study is characterized by the special operating conditions. The high number of material references, the absence of demand patterns, and the high variation of the spare parts orders were the main challenges faced during the project.

The critical activities of the process were identified, and the times invested in each task together with the value-added in the affected areas were assessed. Therefore, we carried out a complete evaluation of the value-added chain of the order processing procedure in the project. This was extremely important to define the key activities to eliminate, improve, or redefine with the ultimate target of increasing the value-added of the complete supply chain. Kaizen ideas were detailed to illustrate the solution strategy proposed for this project. The development of a lean logistics concept, eliminating the waste and increasing the added value of the spare parts supply process, enables the improvement in the delivery fulfillment of the ABC organization to the requesting military units, qualifying the correct achievement of their missions in national territory or operation zone.

The implementation of the two pilot phases of the ideal VSD process performed successfully showing a clear improvement in the key performance indicators. The validation of the actions derived from VSM/VSD in the test period was decisive to finally decide the implementation in the complete order processing procedure from the beginning of 2017. After the ideal VSD process deployment, the results indicated that the future state map could increase added-value activities from 44% to 70%, and the average and deviation of the lead-time was reduced up to 69.6% and 61.9%, respectively, from 2014 to 2017.

The integration of analytical tools to evaluate the system variation, including modeling and simulation of the system before and after the value stream analysis application is recommended as future activities and research lines. The implementation of the lean management approach presented in this work in a military logistics procedure highlights the need for reinforcing these practices in the military context. According to the obtained results, we can also conclude that lean methodologies could be further extended to other military logistics processes and units with the ultimate target of improving the military unit's delivery fulfillment.

Author Contributions: Conceptualization R.A., M.T.; Validation: R.P.-M., J.A.P.; Methodology, R.A., M.T.; Investigation, R.A., M.T., R.P.-M., J.A.P.; Resources, R.P.-M., J.A.P.; Writing—Original Draft Preparation, R.A., M.T.; Writing—Review and Editing, R.P.-M., J.A.P. All authors have read and agreed to the published version of the manuscript.

Funding: This research was funded by the Centro Universitario de la Defensa (Zaragoza-Spain) via project number CUD 2015-21 and by Aragon Government through the Research Activity Grant for research groups.

Acknowledgments: The authors thank the ABC military organization unit for providing the case study and their collaboration.

Conflicts of Interest: The authors declare no conflict of interest.

References

1. The Wharton School, University of Pennsylvania. Managing Supply Chains: What the Military Can Teach Business (and Vice Versa). Available online: http://knowledge.wharton.upenn.edu/article/managing-supply-chains-what-the-military-can-teach-business-and-vice-versa/ (accessed on 17 July 2017).
2. Lai, E.M. An Analysis of the Department of Defense Supply Chain: Potential Applications of the Auto-ID Center Technology to Improve Effectiveness. Ph.D. Thesis, Department of Mechanical Engineering, Massachusetts Institute of Technology, Cambridge, MA, USA, 2003.
3. Joint Vision 2020. In *America's Military—Preparing for Tomorrow*; US Government Printing Office: Washington, DC, USA, 2000.
4. Haraburda, S.; Col, B. Transforming military support processes from logistics to supply chain management. *Army Sustain.* **2016**, *48*, 12–15.
5. Stewart, G. Supply-chain operations reference model (SCOR): The first cross-industry framework for integrated supply-chain management. *Logist. Inf. Manag.* **1997**, *10*, 62–67. [CrossRef]
6. Huan, S.H.; Sheoran, S.K.; Wang, G. A review and analysis of supply chain operations reference (SCOR) model. *Supply Chain Manag. Int. J.* **2004**, *9*, 23–29. [CrossRef]
7. Muckstadt, J.A.; General, T.; Babbitt, G. *Military Supply Chains: Five Principles for Managing Uncertainty*; PTC Inc.: Boston, MA, USA, 2016.
8. Womack, J.P.; Jones, D.T.; Roos, D. *The Machine that Changed the World: The Story of Lean Production*; Harper Perennial: New York, NY, USA, 1991.
9. Liker, J.K. *The Toyota Way: 14 Management Principles from the World's Greatest Manufacturer*; McGraw Hill: New York, NY, USA, 2004.
10. Ohno, T. *Toyota Production System: Beyond Large-Scale Production*; Productivity Press: New York, NY, USA, 1988.
11. Shingo, S. *Non-Stock Production: The Shingo System of Continuous Improvement*; Productivity Press: Cambridge, MA, USA, 1988.
12. Burt, D.N.; Dobler, D.W.; Starling, S.L. *World Class Supply Management: The Key to Supply Chain Management*; McGraw-Hill/Irwin: New York, NY, USA, 2003; ISBN 0072290706.
13. Lysons, K.; Farrington, B. *Purchasing and Supply Chain Management*, 8th ed.; Pearson Education: London, UK, 2012.
14. O'Rourke, D. The science of sustainable supply chains. *Science* **2014**, *344*, 1124–1127. [CrossRef]
15. Lamming, R. *Beyond Partnership: Strategies for Innovation and Lean Supply*; Prentice Hall: Upper Saddle River, NJ, USA, 1993.
16. Lamming, R. Squaring lean supply with supply chain management. *Int. J. Oper. Prod. Manag.* **1996**, *16*, 183–196. [CrossRef]
17. Pande, P.S.; Neuman, R.P.; Cavanagh, R.R. *The Six Sigma Way*; McGraw-Hill: New York, NY, USA, 2000.
18. Imai, M. *Kaizen: The Key to Japan's Competitive Success*; McGraw-Hill Education: New York, NY, USA, 1986.
19. Harry, M.; Schroeder, R.R. *Six Sigma: The Breakthrough Management Strategy Revolutionizing the World's Top Corporations*; Doubleday Business: New York, NY, USA, 1999.
20. Arnheiter, E.D.; Maleyeff, J. The integration of lean management and Six Sigma. *TQM Mag.* **2005**, *17*, 5–18. [CrossRef]
21. Cherrafi, A.; Elfezazi, S.; Chiarini, A.; Mokhlis, A.; Benhida, K. The integration of lean manufacturing, Six Sigma and sustainability: A literature review and future research directions for developing a specific model. *J. Clean. Prod.* **2016**, *139*, 828–846. [CrossRef]
22. Tuttle, W., Jr. *Defense Logistics for the 21st Century*; Naval Institute Press: Annapolis, MD, USA, 2013.
23. NATO Standardization Agency. *NATO Glossary of Terms and Definitions*; CreateSpace Independent Publishing Platform: Scotts Valley, CA, USA, 2013.
24. Zhou, W.; Zhang, C.; Wang, Q. Concealment measurement and flow distribution of military supply transportation: A double-entropy model. *Eur. J. Oper. Res.* **2018**, *264*, 570–581. [CrossRef]
25. Kalchschmidt, M.; Zotteri, G.; Verganti, R. Inventory management in a multi-echelon spare parts supply chain. *Int. J. Prod. Econ.* **2003**, *81*, 397–413. [CrossRef]
26. Huiskonen, J. Maintenance spare parts logistics: Special characteristics and strategic choices. *Int. J. Prod. Econ.* **2001**, *71*, 125–133. [CrossRef]

27. Tysseland, B.E. Spare parts optimization process and results: OPUS10 cases in the Norwegian Defence. *Int. J. Phys. Distrib. Logist. Manag.* **2009**, *39*, 8–27. [CrossRef]
28. Gholz, E.; James, A.D.; Speller, T.H. The second face of systems integration: An empirical analysis of supply chains to complex product systems. *Res. Policy* **2018**, *47*, 1478–1494. [CrossRef]
29. Mayr, A.; Weigelt, M.; Kühl, A.; Grimm, S.; Erll, A.; Potzel, M.; Franke, J. Lean 4.0—A conceptual conjunction of lean management and Industry 4.0. *Procedia CIRP* **2018**, *72*, 622–628. [CrossRef]
30. Krafcik, J.F. Triumph of the lean production system. *MIT Sloan Manag. Rev.* **1988**, *30*, 41.
31. Shah, R.; Ward, P.T. Lean manufacturing: Context, practice bundles, and performance. *J. Oper. Manag.* **2003**, *21*, 129–149. [CrossRef]
32. Bhamu, J.; Sangwan, K.S.; Singh Sangwan, K. Lean manufacturing: Literature review and research issues. *Int. J. Oper. Prod. Manag.* **2014**, *34*, 876–940. [CrossRef]
33. Sundar, R.; Balaji, A.N.; Kumar, R.M.S. A Review on Lean Manufacturing Implementation Techniques. *Procedia Eng.* **2014**, *97*, 1875–1885. [CrossRef]
34. Netland, T.H.; Schloetzer, J.D.; Ferdows, K. Implementing corporate lean programs: The effect of management control practices. *J. Oper. Manag.* **2015**, *36*, 90–102. [CrossRef]
35. Pearce, A.; Pons, D. Advancing lean management: The missing quantitative approach. *Oper. Res. Perspect.* **2019**, *6*, 100114. [CrossRef]
36. Wahab, A.N.A.; Mukhtar, M.; Sulaiman, R. A Conceptual Model of Lean Manufacturing Dimensions. *Procedia Technol.* **2013**, *11*, 1292–1298. [CrossRef]
37. Hines, P.; Holweg, M.; Rich, N. Learning to evolve. *Int. J. Oper. Prod. Manag.* **2004**, *24*, 994–1011. [CrossRef]
38. Rother, M.; Shook, J. *Learning to See: Value Stream Mapping to Add Value and Eliminate Muda*; The Lean Enterprise Institute: Brookline, MA, USA, 1998.
39. Bruce, M.; Daly, L.; Towers, N. Lean or agile: A solution for supply chain management in the textiles and clothing industry? *Int. J. Oper. Prod. Manag.* **2004**, *24*, 151–170. [CrossRef]
40. Christopher, M. The Agile Supply Chain. *Ind. Mark. Manag.* **2000**, *29*, 37–44. [CrossRef]
41. Naylor, J.B.; Naim, M.; Berry, D. Leagility: Integrating the lean and agile manufacturing in the total supply chain. *Int. J. Prod. Econ.* **1999**, *62*, 107–118. [CrossRef]
42. Martin, C.; Towill, D.R. Supply chain migration from lean and functional to agile and customised. *Supply Chain Manag. Int. J.* **2000**, *5*, 206–213. [CrossRef]
43. Minculete, G.; Olar, P. Push and Pull systems in supply chain management. Correlative approaches in the military field. *J. Def. Resour. Manag.* **2016**, *7*, 165–172.
44. Bean, W.L.; Joubert, J.W.; Luhandjula, M.K. Inventory management under uncertainty: A military application. *Comput. Ind. Eng.* **2016**, *96*, 96–107. [CrossRef]
45. Hines, P.; Rich, N. The seven value stream mapping tools. *Int. J. Oper. Prod. Manag.* **1997**, *17*, 46–64. [CrossRef]
46. Forno, A.J.D.; Pereira, F.A.; Forcellini, F.A.; Kipper, L.M. Value Stream Mapping: A study about the problems and challenges found in the literature from the past 15 years about application of Lean tools. *Int. J. Adv. Manuf. Technol.* **2014**, *72*, 779–790. [CrossRef]
47. Singh, B.; Garg, S.K.; Sharma, S.K. Value stream mapping: Literature review and implications for Indian industry. *Int. J. Adv. Manuf. Technol.* **2011**, *53*, 799–809. [CrossRef]
48. Shou, W.; Wang, J.; Wu, P.; Wang, X.; Chong, H.-Y. A cross-sector review on the use of value stream mapping. *Int. J. Prod. Res.* **2017**, *55*, 3906–3928. [CrossRef]
49. Seth, D.; Seth, N.; Dhariwal, P. Application of value stream mapping (VSM) for lean and cycle time reduction in complex production environments: A case study. *Prod. Plan. Control* **2017**, *28*, 398–419. [CrossRef]
50. Wee, H.M.; Wu, S. Lean supply chain and its effect on product cost and quality: A case study on Ford Motor Company. *Supply Chain Manag. Int. J.* **2009**, *14*, 335–341. [CrossRef]
51. Rahani, A.R.; Al-Ashraf, M. Production Flow Analysis through Value Stream Mapping: A Lean Manufacturing Process Case Study. *Procedia Eng.* **2012**, *41*, 1727–1734. [CrossRef]
52. Andrade, P.F.; Pereira, V.G.; Del Conte, E.G. Value stream mapping and lean simulation: A case study in automotive company. *Int. J. Adv. Manuf. Technol.* **2016**, *85*, 547–555. [CrossRef]
53. Venkataraman, K.; Ramnath, B.V.; Kumar, V.M.; Elanchezhian, C. Application of Value Stream Mapping for Reduction of Cycle Time in a Machining Process. *Procedia Mater. Sci.* **2014**, *6*, 1187–1196. [CrossRef]

54. Rohac, T.; Januska, M. Value Stream Mapping Demonstration on Real Case Study. *Procedia Eng.* **2015**, *100*, 520–529. [CrossRef]

55. Hodge, G.L.; Ross, K.G.; Joines, J.A.; Thoney, K. Adapting lean manufacturing principles to the textile industry. *Prod. Plan. Control* **2011**, *22*, 237–247. [CrossRef]

56. Abdulmalek, F.A.; Rajgopal, J. Analyzing the benefits of lean manufacturing and value stream mapping via simulation: A process sector case study. *Int. J. Prod. Econ.* **2007**, *107*, 223–236. [CrossRef]

57. Rohani, J.M.; Zahraee, S.M. Production Line Analysis via Value Stream Mapping: A Lean Manufacturing Process of Color Industry. *Procedia Manuf.* **2015**, *2*, 6–10. [CrossRef]

58. Yang, T.; Kuo, Y.; Su, C.T.; Hou, C.L. Lean production system design for fishing net manufacturing using lean principles and simulation optimization. *J. Manuf. Syst.* **2015**, *34*, 66–73. [CrossRef]

59. Hartmann, L.; Meudt, T.; Seifermann, S.; Metternich, J. Value stream method 4.0: Holistic method to analyse and design value streams in the digital age. *Procedia CIRP* **2018**, *78*, 249–254. [CrossRef]

60. Knoll, D.; Reinhart, G.; Prüglmeier, M. Enabling value stream mapping for internal logistics using multidimensional process mining. *Expert Syst. Appl.* **2019**, *124*, 130–142. [CrossRef]

61. Stadnicka, D.; Litwin, P. Value stream mapping and system dynamics integration for manufacturing line modelling and analysis. *Int. J. Prod. Econ.* **2019**, *208*, 400–411. [CrossRef]

62. Jiménez, E.; Tejeda, A.; Pérez, M.; Blanco, J.; Martínez, E. Applicability of lean production with VSM to the Rioja wine sector. *Int. J. Prod. Res.* **2012**, *50*, 1890–1904. [CrossRef]

63. De Steur, H.; Wesana, J.; Dora, M.K.; Pearce, D.; Gellynck, X. Applying Value Stream Mapping to reduce food losses and wastes in supply chains: A systematic review. *Waste Manag.* **2016**, *58*, 359–368. [CrossRef]

64. Tyagi, S.; Choudhary, A.; Cai, X.; Yang, K. Value stream mapping to reduce the lead-time of a product development process. *Int. J. Prod. Econ.* **2015**, *160*, 202–212. [CrossRef]

65. Kaiser, J.; Urnauer, C.; Metternich, J. A framework for planning logistical alternatives in value stream design. *Procedia CIRP* **2019**, *81*, 180–185. [CrossRef]

66. Ali, N.B.; Petersen, K.; De França, B.B.N. Evaluation of simulation-assisted value stream mapping for software product development: Two industrial cases. *Inf. Softw. Technol.* **2015**, *68*, 45–61. [CrossRef]

67. Grove, A.L.; Meredith, J.O.; Macintyre, M.; Angelis, J.; Neailey, K. Lean implementation in primary care health visiting services in National Health Service UK. *Qual. Saf. Health Care* **2010**, *19*, e43. [CrossRef]

68. Garza-Reyes, J.A.; Torres Romero, J.; Govindan, K.; Cherrafi, A.; Ramanathan, U. A PDCA-based approach to Environmental Value Stream Mapping (E-VSM). *J. Clean. Prod.* **2018**, *180*, 335–348. [CrossRef]

69. Faulkner, W.; Badurdeen, F. Sustainable Value Stream Mapping (Sus-VSM): Methodology to visualize and assess manufacturing sustainability performance. *J. Clean. Prod.* **2014**, *85*, 8–18. [CrossRef]

70. Helleno, A.L.; de Moraes, A.J.I.; Simon, A.T. Integrating sustainability indicators and Lean Manufacturing to assess manufacturing processes: Application case studies in Brazilian industry. *J. Clean. Prod.* **2017**, *153*, 405–416. [CrossRef]

71. Deshkar, A.; Kamle, S.; Giri, J.; Korde, V. Design and evaluation of a Lean Manufacturing framework using Value Stream Mapping (VSM) for a plastic bag manufacturing unit. *Mater. Today Proc.* **2018**, *5*, 7668–7677. [CrossRef]

72. Kasava, N.K.; Yusof, N.M.; Khademi, A.; Saman, M.Z.M. Sustainable domain value stream mapping (SdVSM) framework application in aircraft maintenance: A case study. *Procedia CIRP* **2015**, *26*, 418–423. [CrossRef]

73. Vinodh, S.; Ben Ruben, R.; Asokan, P. Life cycle assessment integrated value stream mapping framework to ensure sustainable manufacturing: A case study. *Clean Technol. Environ. Policy* **2016**, *18*, 279–295. [CrossRef]

74. Edtmayr, T.; Sunk, A.; Sihn, W. An Approach to Integrate Parameters and Indicators of Sustainability Management into Value Stream Mapping. *Procedia CIRP* **2016**, *41*, 289–294. [CrossRef]

75. Haefner, B.; Kraemer, A.; Stauss, T.; Lanza, G. Quality Value Stream Mapping. *Procedia CIRP* **2014**, *17*, 254–259. [CrossRef]

76. Toivonen, T.; Siitonen, J. Value Stream Analysis for Complex Processes and Systems. *Procedia CIRP* **2016**, *39*, 9–15. [CrossRef]

77. Oberhausen, C.; Plapper, P. Cross-enterprise value stream assessment. *J. Adv. Manag. Res.* **2017**, *14*, 182–193. [CrossRef]

78. Roh, P.; Kunz, A.; Wegener, K. Information stream mapping: Mapping, analysing and improving the efficiency of information streams in manufacturing value streams. *CIRP J. Manuf. Sci. Technol.* **2019**, *25*, 1–13. [CrossRef]

79. de Mast, J.; Lokkerbol, J. An analysis of the Six Sigma DMAIC method from the perspective of problem solving. *Int. J. Prod. Econ.* **2012**, *139*, 604–614. [CrossRef]

80. Parast, M.M. The effect of Six Sigma projects on innovation and firm performance. *Int. J. Proj. Manag.* **2011**, *29*, 45–55. [CrossRef]

81. Srinivasan, K.; Muthu, S.; Devadasan, S.R.; Sugumaran, C. Enhancing effectiveness of shell and tube heat exchanger through six sigma DMAIC phases. *Procedia Eng.* **2014**, *97*, 2064–2071. [CrossRef]

82. Nithyanandam, G.K.; Pezhinkattil, R. A Six Sigma approach for precision machining in milling. *Procedia Eng.* **2014**, *97*, 1474–1488. [CrossRef]

83. Cunha, C.; Dominguez, C. A DMAIC Project to Improve Warranty Billing's Operations: A Case Study in a Portuguese Car Dealer. *Procedia Comput. Sci.* **2015**, *64*, 885–893. [CrossRef]

84. Erbiyik, H.; Saru, M. Six Sigma Implementations in Supply Chain: An Application for an Automotive Subsidiary Industry in Bursa in Turkey. *Procedia Soc. Behav. Sci.* **2015**, *195*, 2556–2565. [CrossRef]

85. Hardy, D. *An Analysis of Lean Six Sigma in the Army, Navy and Air Force*; Naval Postgraduate School: Monterey, CA, USA, 2018.

86. Baily, A.; Gibson, B.; MacFarlane, J.M.; Hazelbaker, C.; Richards, J.; Lyle, I.; Rawhouser, B. Application of lean six sigma to reduce repeated handling of material at Tobyhanna Army Depot. In Proceedings of the 39th International Annual Conference American Society Engineering Management, ASEM 2018 Bridging Gap between Engineering and Business, Coeur d'Alene, ID, USA, 17–20 October 2018; pp. 691–700.

Article

Development of a Pull Production Control Method for ETO Companies and Simulation for the Metallurgical Industry

Javier Gejo García, Sergio Gallego-García and Manuel García-García *

Department of Construction and Fabrication Engineering, National Distance Education University (UNED), 28040 Madrid, Spain; j_gejo@yahoo.es (J.G.G.); sgallego118@alumno.uned.es (S.G.-G.)
* Correspondence: mggarcia@ind.uned.es; Tel.: +34-682-880-591

Received: 29 October 2019; Accepted: 27 December 2019; Published: 30 December 2019

Abstract: At the moment, many engineer-to-order manufacturers are under pressure, the overcapacity in many sectors erodes prices and many companies, especially in Europe have gone into recent years in bankruptcy. Due to the increasing competition as well as the new customer requirements, the internal processes of an ETO company play an essential role in order to achieve a unique selling proposition (USP). Therefore this paper exposes how the production planning and control of an engineer-to-order manufacturer can be designed in order to increase its OTD (order-to-delivery) rate as well as decrease the WIP (work-in-progress) and the production lead times. To prove the optimized planning logic, it was applied in a simulation case study and based on the results; the conclusions about its potential are derived.

Keywords: engineer-to-order; order management process; production planning and control; capacity planning; maintenance management; order-to-delivery; agent-based simulation; metallurgical industry

---

1. Introduction

Markets for engineer-to-order (ETO) manufacturers that were stable in the past are now dynamic and uncertain in which prices have reduced over the last decades [1] (p. 43). At the moment, many ETO manufacturers, such as companies in the metallurgical sector, working for specific needs and end-segments such as the steel industry, are under pressure; the current overcapacity is destroying prices and profitability [2] (p. 2) and many producers have closed facilities, have made layoffs and have gone in bankruptcy in recent years [3] (pp. 1–2). As an example by the end of 2015, U.S. steel producers were utilizing less than 65% of their capacity, and were forced to lay off 12,000 employees in this year [3] (p. 1). In 2007–2011 capacity utilization was about 80% and supply and demand was in a balanced state, but since 2012 the overcapacity challenge was becoming obvious, with 74.8% in 2014 [4] (p. 62), less than 70% in 2015 [3] (p. 5) and with a recovery in 2018 due to a reduction of capacity since 2015 with 75.7% capacity utilization with 2233.7 million tons of capacity available [5] and 1690.1 of produced crude steel [6] (p. 2). The overcapacity created was driven by new investment from old market leaders as well as new emerging market companies in the last decades like in China, which now has two-thirds of world steel overcapacity [3] (p. 1).

On the other hand, in respond to that, ETO companies are continuously seeking for ways or methods how to reduce costs and lead times as well as increasing their external flexibility [1] (p. 43). In the literature there is limited research related to supply chain management in the low-volume engineer to order (ETO) sector, in contrast to the extensive literature on the high-volume sector, particularly automotive and electronics [7] (p. 179). The limited research that has been undertaken in the low volume ETO sector has focused on production control, information systems, manufacturing systems and the co-ordination of marketing and manufacturing [1] (p. 44).

In this global market situation, current producers need to work on their technological and organizational advantage. On the one side, the steel industry has a low frequency of innovation compared to other industries with process technologies have been used for decades [8] (p. 8). Until the recession in 1973, advanced countries such as US, Japan and European countries were increasing their steel production [8] (p. 9). Due to the recession, steel demand decreased and US and Japanese steel firms were forced to export their technology [8] (p. 10). Therefore, the first potential unique selling proposition (USP), the technological advantage, hardly exists anymore for US, Japanese, or European countries, as many companies worldwide have access and have invested in new equipment in recent decades and years, mainly in developing countries [9] (pp. 2–3) such as Korea, Taiwan, Brazil, and China [8] (p. 10).

For ETO companies, the question arises of what options remain to improve their competitive situation and to shape a long-term and profitable business model. In addition, additional cost savings programs will not be enough to achieve sustainable returns. For this reason, ETO producers need to be focus on the second potential USP, the internal process optimization, such as digitization and optimization of business processes enabling a better service level to end-customers.

1.1. Research Questions and Goals

On the basis of the previous paragraphs, this paper aims to provide a methodology for the improvement of internal processes in regard to production planning and control by optimizing internal parameters, work-in-progress (WIP), and production lead times, as well as end-customer service level, and order-to-delivery (OTD). The final goal of this optimization would be to increase the company's competitiveness securing its long-term existence. Based on these objectives the following research questions can be declared:

- Which are current challenges of ETO manufacturers?
- Which is the current common production planning strategy being used in ETO industries?
- How do other production planning strategies can be applied to the ETO sector?
- What is the potential benefit of this change?
- In order to achieve these goals, this paper presents the following structure:

    1. Introduction: definition of challenge, objectives, and simulation.
    2. Literature review of:

        ○ Engineer-to-order typology and characteristics;
        ○ Organizational model for the order management process;
        ○ Production planning and control: push versus pull;
        ○ Capacity planning and maintenance;
        ○ Theory of constraints (TOC) and drum-buffer-rope (DBR);
        ○ Agent-based simulation.

    3. Typical initial situation and development of pull production control for ETO manufacturers.
    4. Applying the model for production control to the metallurgical industry through simulation.
    5. Simulation results and implementation.
    6. Conclusions.

The main expected outcome of the research paper is that a pull control approach using drum-buffer-rope (DBR) approach will lead to an improvement of OTD and to a reduction of lead times and WIP stocks.

1.2. Methodology Used

The research methodology for an ETO company consists of the steps shown in Figure 1:

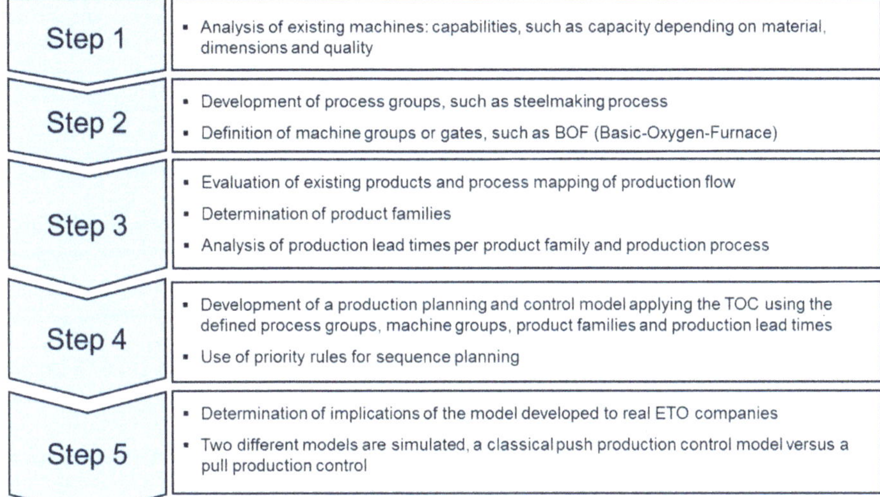

Figure 1. Methodology used: steps.

Based on the previously described methodology two different models are simulated, a classical push production control model versus a pull production control. The main objectives are: to generate knowledge of the supply chain, development and validation of improvements using what would happen if of analysis and quantification of the benefits of decision support at the level of strategic decision making [10] (p. 4). One recent modeling method is agent-based simulation and it has no standard language. The structure of an agent based model is created using graphical editors depending on the software. The behavior of agents is specified in many different ways. Frequently, the agent has a notion of state, and its actions and reactions depend on its state [11] (p. 14). For the research work AnyLogic was used and the models presented production orders as agents with their own characteristics. As a consequence the research was qualitative in the conceptualization and quantitative for the simulation models and data used. In Figure 2 it can be shown the steps followed to perform the simulation of the two models:

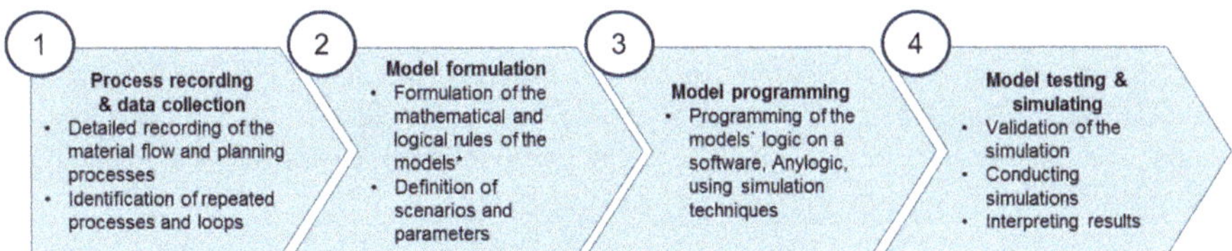

Figure 2. Simulation methodology used.

2. Theoretical Background as the Basis for Model Development

2.1. Engineer-To-Order Typology and Characteristics

Based on the literature six different supply chain typologies can be defined to describe the range of possible operations: engineer-to-order (ETO), buy-to-order (BTO), make-to-order (MTO), assemble-to-order (ATO), make-to-stock (MTS), and ship-to-stock (STS). The ETO supply chain is described as a supply chain where the customer order decoupling point is located at the design stage, so each customer order influenced the design phase of a product [12] (p. 741).

The characteristics of the manufacturers with ETO are customized and high-value added products, in low volume with deep and complex product structure in order to meet specific customer requirements [1,7] (p. 43, p. 179). It is primarily associated with large, complex project environments in sectors such as construction and capital goods. However, while the term ETO is used, in the literature exists confusion about the appropriate strategies and there is no major systematic literature reviews or syntheses of knowledge relating to the ETO supply chain type [12] (p. 741).

2.2. Competitiveness

The competitive advantage of ETO companies is based on the fulfillment of individual customer requirements [13] (p. 16). This approach demands a high flexibility of the manufacturing process [14] (p. 1). The success factor is the capability to deliver the specific order in time [15] (p. 384). As a consequence, a short lead time and effective synchronization of the production management processes are key [14] (pp. 1–2) to gain competitiveness.

Between innovation and competitive advantage, there is a complex and multidimensional relationship [16,17], in which sustainability plays a primary role. Therefore, the innovation must be characterized by sustainability [18] (p. 8).

As an example, the competitiveness of steel production is based on a set of internal comparative advantages such as labor costs, consistent economy and foreign trade policy, long-term work experience, energy prices, or highly efficient work organization [19] (p. 122). These indicators provide us how these factors differ and how they affect the international competitiveness of steel products [19] (p. 122). Based on that, the EU is losing its competitiveness in those market segments where it is not possible to apply other competitive advantages such us: innovation, know-how, or scientific and research outputs for commercial purposes [19] (p. 120).

2.3. Organizational Model for the ETO's Order Management Process

There are three stages of interaction between ETO companies and their customers. The first is marketing [7] (p. 188). The second stage is tendering that involves the preliminary development of the conceptual design and the definition of major components and systems. A technical specification, delivery schedule, price, and commercial terms are agreed. Of costs 75–80% are committed at this stage. The third stage takes place after a contract has been awarded and includes non-physical processes, such as design and planning, and physical processes associated with manufacturing, assembly and commissioning. Supply chain management in ETO companies involves the co-ordination of internal processes across these three stages [7] (p. 188)

ETO companies span a continuum from a fully integrated company that manufactures all components and assemblies at one extreme, to a pure design and contract organization at the other. The appropriate structure for a particular company is dependent upon many factors including cost, capital available for equipment, potential utilization of plant, internal and external capabilities and flexibility. These factors vary from firm to firm giving rise to different levels of vertical integration [7] (p. 188).

2.4. Production Planning and Control: Push Versus Pull

One of the central tasks of production management is production planning [20] (p. 27). The original tasks of production planning include the planning of the products to be manufactured, as well as the required production factors and processes [21] (p. 29). Production management contains the tasks of design, planning, monitoring, and control of the productive system and business resources such as people, machines, material, and information [22] (pp. 249–273). In this context the task of production planning and control is the planning and control of deadlines, delivery dates, capacities and quantities of manufacturing, and assembly processes [21] (p. 29).

The planning of production processes contains a comprehensive planning of delivery dates oriented to customer orders, a reservation planning of the capacities oriented to each of the machines (sequence planning) and a planning of the personnel and material to be available [23].

The processes in a supply chain are divided into two categories depending on whether they are executed in response to a customer order or in anticipation of customer orders. Production according to the "pull" principle is initiated by customer orders, while the "push" principle is initiated and carried out in anticipation of customer orders and is usually based on demand forecasts. This dilemma is extremely useful when considered the strategic decisions related to the design of the supply chain, and therefore of the production [24] (p. 14). The objective of the design of a productive system within a supply chain is to find the balance between the two. Balance is described by the decoupling point of the customer order [25] (p. 52). An ETO manufacturer is a producer following mainly the pull principle across the supply chain with the decoupling point at the beginning of it or even not existing well pure pull is applied.

In terms of the publication the concept of push and pull compared does not correspond to the nature of producing in response or anticipation of customer orders, but it deals with the regulation and control method applied in the production process in order to release orders into production in the different steps. A push control approach initiates production trying to maximize production utilization without considering stocks and the restrictions along the production process. On the other hand a pull approach is based on the TOC and considers the restrictions and stocks in the current and future state of the production system.

2.5. Capacity Planning and Maintenance

In this research paper it was considered the core capacity planning tasks and its relationship with maintenance management. First, based on sales information the production master program (PMP) determines what products should be produced and in what quantities at the following planning intervals [26] (p. 57). The PMP is subdivided into partial processes of sales planning, primary gross needs planning, primary net requirements planning, and approximate resource planning [21] (pp. 41–43). The result of the PMP is a coordinated production program when considering the productive capacities and sales expectations of the company [21] (p. 185). In the approximate planning of the resources it is verified if the sales planning and the production programs can be carried out with the available resources according to the type, quantity, and date on which they are planned and compared with available resources. In this context, resources are personnel, facilities, tools and material [21] (p. 43).

Process programming provides temporary relationships between production orders. There are three methods: progressive programming, regressive programming, and midpoint programming. The programming of the process can be carried out considering finite or infinite capacities. Capacity is the output of an installation for a period of time. Capacity calculation for a production system depends on maintenance in regard of machine availability. Moreover to calculate the total output of the production system the quality and performance rates have to be added. For ETO manufacturers capacity planning vary normally a lot depending on the product type due to the differences in terms of time needed.

The capacity demand is then compared with the supply of available capacity. Based on this there are two options in case the required capacity exceeds the offer. The first is to adjust the capacity by increasing it thanks to overtime or special shifts. The second is to postpone demand peaks by shifting manufacturing orders to later points in time [21] (pp. 48–50).

Sequencing of orders is carried out later with the help of criteria selected as priority rules or minimization of equipment preparation times. The release of production orders is carried out taking into account detailed resource planning and detailed programming. In doing so, defined release rules or methods are used, such as load-oriented order release [21] (pp. 55–57).

2.6. Theory of Constraints (TOC) and Drum-Buffer-Rope (DBR)

The theory of constraints (TOC) suggests the application of demand-pull approach combined with buffer management to effectively manage inventory [27] (p. 23). The initial target buffer size, when to adjust the buffer and the quantity to be adjusted are key management decisions that determine whether the demand-pull approach can be successfully applied in practice or not [27] (p. 23).

DBR is the operational production planning and control approach within the theory of constraints (TOC) introduced by Goldratt (1993). Based on it Gupta et al. (2002) investigated the workings of TOC with the help of simulation. The main principle is to subordinate the production plan to the system's capacity-constrained resource (CCR). Buffers secure the CCR and the finished goods inventory against starvation. The buffers are forecast of processing and transfer times plus a certain amount of safety time. The area between material release point and CCR is covered by the CCR buffer, while the area between CCR and the customers is controlled through the shipping buffer [28] (p. 2182).

3. Typical Initial Situation and Development of Pull Production Control for ETO Manufacturers

The production project pursued the following objectives, the development of pull planning method for the improvement of the OTD (order-to-delivery), the reduction of the stock in WIP and the reduction of the production lead times. The methodological steps are:

- At the beginning of the project a process recording has to take place. It can be done using value stream mapping (VSM) or commercial tools for process mapping such as Microsoft Visio or Airis.
- On the basis of the analysis of the current process, the challenges can be normally identified. Typical internal company challenges are:

  - No overarching coordination of delivery dates between sales and production;
  - Bottlenecks considered mainly in a reactive approach;
  - Complex process with different production types along the production process;
  - Current production planning and control strategy: normally a classic push control; where the operative production areas always have to keep running, whether it is necessary or not;
  - Volume or production unit as example tons of material as goal for operative production managers.

- Analysis of the impact of the facts analyzed. On the basis of the previously described challenges, the typical consequences of those common issues are:

  - Local optimization;
  - Initiation of production without complete customer specifications (blocked inventory);
  - Long lead times;
  - Insufficient delivery reliability;
  - High inventories levels.

- After recording and analysis of the current process and its impact, the development of a new concept has to be initiated. In this paper the so-called "bottleneck control" was selected. The principle of bottleneck control explains that the bottlenecks ultimately determine the performance of the entire production system and thus represent the clocks. If the bottlenecks are known, it is sufficient to plan the bottlenecks. Since bottleneck control also has a regulating effect on the production stock, there are two further positive effects: stock costs are reduced and throughput times are reduced.

The model suggests the implementation of four essential components for the design of the bottleneck control within an organization with ETO manufacturing:

1. Definition of machine groups or gates structures: first of all machines with similar process steps are classified and alternative machines in terms of technical processing and output have to be identified. By summing the total capacity of a gate structure, the output of a production process group per period of time can be determined.
2. Development of product families: to reduce complexity, all sales items that follow the same production path have to be grouped together. To do this, process modules and machine groups or gates have to be determined. This made it possible to link the new production product families with the sales product families. Additionally a list of typical bottlenecks per product family.
3. Determination and updating model of production lead times: lead times for each product family in each process step or machine groups have to be determined: waiting, processing, and transport times have to be settled by analyzing actual data. Lead time determination needs to have a correlation with product families in sales so the promises of sales employees to end-customers can be met. The lead times have to be updated based on real data feedback in a constant frequency period.
4. Production capacity forecast: in order to provide transparency, all machine groups and gates with robust capacity have to be forecasted on a time unit, as an example in an hour basis.
5. Interface to sales: controlled communication process between sales and production to assign realistic delivery dates.

The new pull production planning and control logic also includes two important prerequisites that are shown in Figure 3:

- Each item will have an order release;
- Each item will have a completion date.

Being both centrally controlled by a production planning area would allow global optimization of the overall system as well as optimal utilization of the bottleneck.

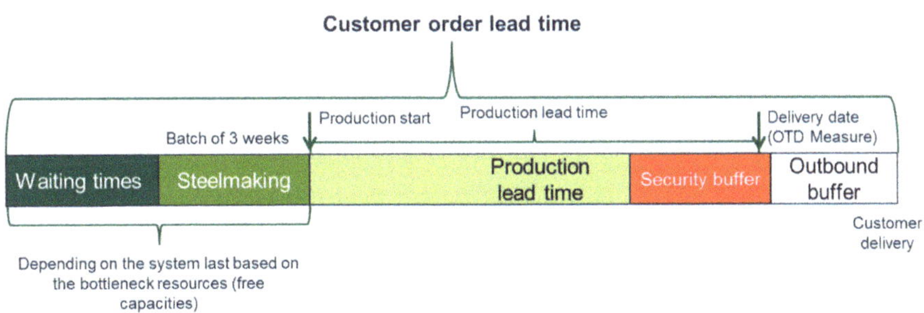

Figure 3. Pull-production control logic.

4. Applying the Model for Production Control to the Metallurgical Industry through Simulation

In order to make the decision and to increase the acceptance for future implementations in ETO manufacturers, an agent-based simulation was carried out in order to show the potential of this approach. The goal of this simulation is the comparison of today's common production control logic, the push control logic, with potential pull control logic.

4.1. Process Recording and Data Collection

As explained the methodology the first step was to represent with a process mapping the material flow of the ETO metallurgical producer and associating the data to the specific processes and product families. The important representatives of the individual product families were depicted in a process mapping tool and then brought into the simulation model with their respective production routes.

For each product family, the needs of five years were taken to design the base scenario. The agents that go through the simulation were the individual manufacturing positions or units with their parameters such as tons, volume, processing time per machine group, order release date, completion date, the transport times between machine groups, etc. In addition, these positions run through the various machine groups along its production. Each machine group includes a number of machines that manufactures the same products. The machine groups depend on their shift model, usually 15 or 21 shifts as explained before.

4.2. Model Formulation

Later, the logical differences between the models are to be described. Both models have the same assumptions as well as the same framework conditions:

- Time horizon: The tasks are assigned according to their temporal relevance at different planning levels. According to the St. Gallen management model, the strategic planning level has a planning horizon of several years [29] (p. 80). Based on it, five years were chosen for the simulation of the two models.
- Production mix: Seven different production families with 12 different production routes were considered.
- Finite capacity for all production steps. Production capacity as the sum of the capacities of the machines within a machine group.
- Processing times depend on the product family, on the variant of product family and on the weight of the production unit.
- Same transport times between buffers and machine independently of the product family.
- Infinite transport capacity between machines along the production process.
- Infinite stock capacity before, along, and at the end of the production process.
- Raw and operating materials are always available for the production process.
- Assumed that in 3 weeks a batch for a steel type can be produced.
- Quality problems or rework lead times are not considered.
- Personal planning not considered.
- Depending on the machine group, 15 or 21 shifts per week.
- Production units are the agents that flow along the production process.
- If the load of a bottleneck is too high, the units wait to get a release date. If there are a certain number of units waiting for it, the manufacturer loses the demand that influencing the bottleneck until the number of units falls below the limit.

Table 1 describes the main differences between both models. The push control starts the production in the steel mill, as far as the orders are in the order intake, since it follows the principle that the machines must not remain still. This causes the production orders to start too early or too late, usually too early, and as a consequence, the WIP increases. In addition, the delivery date determination is not determined on the basis of the bottleneck resources per product family. This, together with a local optimization in the individual production steps, leads to situations with high stock, long lead times and poor on-time delivery.

On the other hand, in the pull control, order release into production is based on the load of the bottleneck resources. For each product family, there are certain typical bottleneck resources to analyze (e.g., indirect stock and direct stock before a machine group converted to processing hours). Indirect stock of a machine group is the stock that has to be processed by this machine group, which is already in production and that is not ready to initiate production in this machine group. By knowing the total amount of processing hours needed to process the direct stock and the indirect stock, the model can determine which one of the machine groups is a bottleneck resource. Therefore, delivery dates are determined by dynamic lead times, which are determined on the basis of the system load. The third difference is the introduction of a global priority rule, which forms the basis for sequencing before each production step. The priority of a position (%, as a percentage) is determined on the basis of the time consumption at the delivery date, the higher the priority the sooner the position has to be processed in that specific production step. This means that at every time period priorities are recalculated to create a list of next production units to be processed in each machine group to optimize the system globally by processing the critical units according to promised delivery date at first.

Table 1. Differences between push and pull production control models for a metallurgical manufacturer.

No.	Difference	Push-Control	Pull-Control Applying TOC/DBR
1	Order release	Order release for improving capacity utilization in the first production steps	Regulated order release control based on the system load
2	Determination of delivery dates	Same order release for units of the same product family	Adjusted based on the system load
3	Sequence planning	First in First out (FIFO)	Global Priority Rule: Priority determined based on time consumption on the delivery date

The two models have moving bottlenecks depending on the product mix that are in a particular moment in production process. The pull-control model can anticipate the bottlenecks by knowing the processing time needed in the current bottleneck resources. Based on this processing time a reliable delivery date is given.

The previously described characteristics are shown in Table 1.

4.3. Model Programming

The simplified production flow simulated can be seen in Figure 4 below:

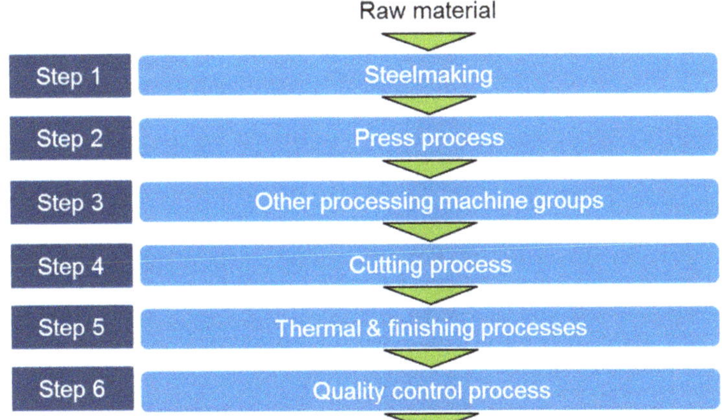

Figure 4. Simplified production flow of the simulation model: 6 main steps.

The models were created using delay times as lead times for production processes, transport, waiting times, etc. The AnyLogic software allows providing a processing time or delaying time depending on the agent entering in a machine group. When an agent or production unit enters a machine, the processing time that depends on the agent is activated. After this time the unit goes out of the machine and waits to be transported to the next processing step.

Moreover, the agents within the simulation are the production units with their parameters associated to them as listed below:

- Product family (number);
- Variant within the product family (number);
- Weight (tons);
- Processing lead time per production step (days);
- Days until security buffer (days);
- Transport lead times between all combinations of production steps (days).

Based on the flow of a production unit along its production process, the following parameters are calculated dynamically within the agents:

- Waiting time until order release is given (days);
- Promised production lead time (days);
- Production lead time (days);
- Transport lead time (days);
- Waiting times along the production process (days);
- Days before or after the promised production lead time (days).

Demand is created within the model with gamma distribution and is equal for both models. Its value is quantified according to the rate of incoming units or positions of a certain product family and variant and it is recalculated every 90 days, assuming that a certain pattern remains for 90 days.

The key performance indicators (KPIs) for the simulation model are:

- Demand: production units ordered (units). The value is written in an Excel file in periods of 90 days.
- Production units started: production units released (units). The value is written in an Excel file in periods of 90 days.
- OTD (on-time-delivery): percentage of units produced before the promised delivery date for logistics (%). The value is written in an Excel file in periods of 90 days.
- Production throughput: cumulated production (tons).
- WIP: quantity of units in production process (units).
- Stock before and after production process: quantity of units before and after the production process (units).

The KPIs per production family are: demand (units), OTD (%), production lead time since order release (days), lead times for technical processing, transport and waiting times (days), percentage of units that did not reach the security buffer (%), and the weight break-down of the units (tons and %).

The KPIs per machine group are: capacity utilization (%), planned load (weeks), priority of a unit before a machine group (%), and direct stock (weeks).

4.4. Model Testing and Simulating

The models can be initiated with WIP production units or without them depending on the adjustable parameters. Moreover, this and other adjustable parameters are shown in Table 2:

Table 2. Adjustable parameters.

No.	Adjustable Parameter	Description	Unit
1	Demand	Expected value and deviation based on gamma distribution	Units per week
2	Maximal load per machine group	Production system load	Days
3	Production lead time	Time for production for the next orders	Days
4	Security buffer	Time for buffer for the next orders	Days
5	Quantity of machines	Number of machines per machine group	Machines
6	Shift model	Determines the planned production time	Shifts per week
7	Performance factor	Considers availability and performance losses	%
8	WIP at time = 0	Quantity of units in production process at day 0	Yes/No

Moreover, the values for the described parameters are to be introduced in the cockpit of the simulation model within the AnyLogic software, with the exception of the shift model that should be introduced in other screen within the model, as depicted in Figure 5:

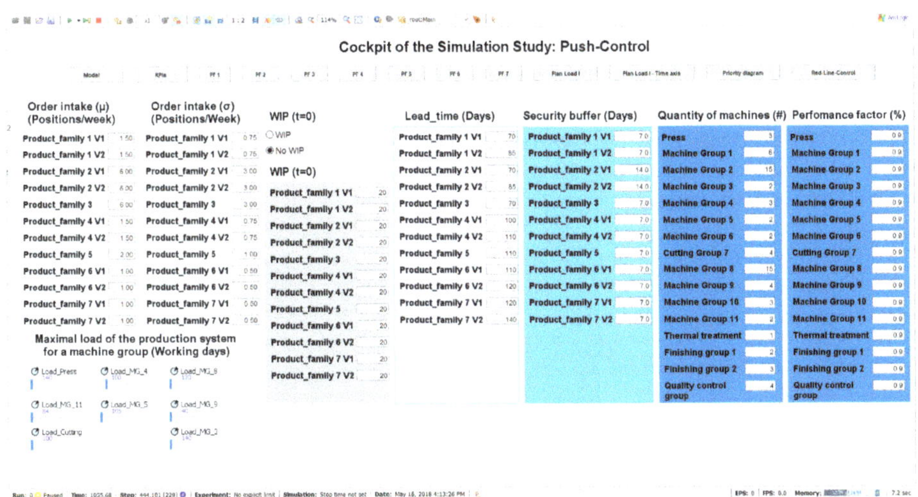

Figure 5. Cockpit of the simulation study for the push-control model.

In addition, Figure 6 shows the testing process performed within the simulation software:

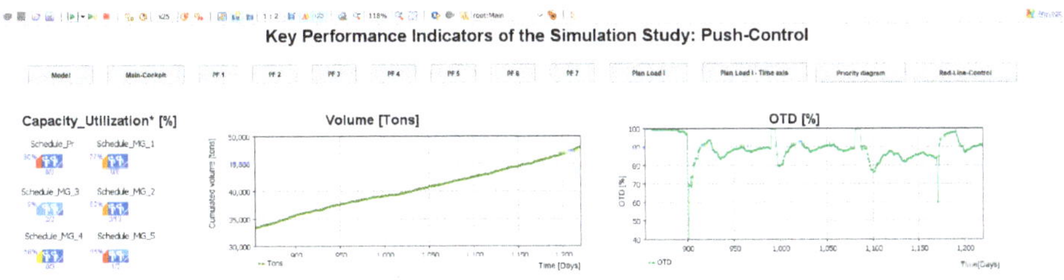

Figure 6. Testing the model with the extreme-value test.

5. Simulation Results and Implementation

The comparison was conducted with the following characteristics:

- Demand created within the model based on statistical distributions that are equal in both models.
- The following demand scenarios were performed are shown below in Table 3:
- Results are written in a "KPI (Key Performance Indicator)" Excel file and then compared between the models.

Table 3. Demand scenarios.

No.	Scenario	Description	Name
1	Scenario 1	Demand in an average level around 30 units per week	Base
2	Scenario 2	High demand for all product families compared to base scenario	High
3	Scenario 3	Low demand for all product families compared to base scenario	Low
4	Scenarios 4–10	High demand for one product family and base for the others	High for product family X

Results for all the scenarios were better for the pull-control with higher difference for scenario 2 and less difference for scenario 3. The results in Table 4 are those for the base scenario, scenario 1, and show a clear competitive advantage for the pull control, as it had achieved better on-time delivery with less inventory and idle time at the same throughput. The biggest difference in the results happened when there was a large order intake. In this case, the pull control was able to smooth the demand with the determination of order release and completion dates, as it could act on the bottleneck early on, but the push-control could only react reactively when it was already too late:

Table 4. Simulation results for push and pull-demand production control—base demand scenario.

No.	Key Performance Indicator	Push-Control	Pull-Control Applying TOC/DBR
1	OTD (%)	74	95
2	Production throughput (tons)	75	77
3	WIP (units)	339	285
4	Waiting time (%) from total production lead time	51	34
5	Capacity utilization (%)	58	61

6. Conclusions

This chapter was divided into theoretical, managerial, and empirical conclusions explaining also the limitations of the research performed as well as describing the potential research work derived:

Theoretical conclusions: the research paper pursued the following objectives, the improvement of the OTD (delivery schedules), the reduction of the stock in WIP and the reduction of the production lead times. In order to achieve these goals, a pull production control with a so-called "bottleneck control" was developed for an ETO manufacturer providing steps as guidelines to follow in order to apply it successfully in real manufacturers.

- A new concept using a TOC approach was developed.
- Steps for the implementation of the new concept developed were described.

Managerial conclusions:

- The current challenges of ETO manufacturers were described and those influencing the production system were integrated in the push control model.
- Internal efficiency was proved as key for metallurgical companies, in particular, for the steel industry.
- The importance of a controlled and synchronized communication between sales and production to assign realistic delivery dates that lead to higher customer satisfaction.

Empirical conclusions: to prove the utility of the new concept a simulation for a metallurgical example was created by including the necessary instruments and control systems developed as well as the validation by means of simulation. The model considered and simulated the variety of complex process such a batch production in the steelworks and unique production items in the other production steps as well as the interrelationships between them. That is why the planning and control system was robust, simple, and transparent to every employee and for the end-customer.

- Both concepts, pull and push-control, were implemented in AnyLogic software with agent-based modeling and simulated to compare the performance in different demand scenarios.
- The benefits of the change from a common push control approach to an approach using DBR are:

  - Global optimization of the production system;
  - Initiation of production with complete customer specifications;
  - Shorter lead times;
  - Higher delivery reliability;
  - Lower inventories levels.

Limitations of the research work:

- Assumed that existing ETO producers used a production control based on a push approach.
- Complexity of an ETO manufacturer was partially built in the simulation model.
- Complexity of processes such as for steelmaking was not built in detail.
- Organization structure and interfaces were not considered in the simulation model.
- Quality problems were not considered.
- Concept was not proved in any company.

Future research: the potential research derived from this paper is:

- Transfer this research method to real production systems applying it in particular cases as an assistance tool for sales and production planning leaders and controllers by centralizing all data related to a topic in a short period of time enabling simulation of what-if-scenarios.
- Consider organization units and their communication within the simulation model.
- Improve the model from implementation feedbacks as well as apply it for production networks with several production plants.

To sum-up, the research work show the potential benefits of a capacity management based on a bottleneck control approach in which committed delivery dates can be met as well as improving internal production key performance indicators leading to an increase in competitiveness.

Author Contributions: Conceptualization, M.G.-G., J.G.G. and S.G.-G.; Methodology, M.G.-G. and S.G.-G.; Software and programming, S.G.-G.; Validation, J.G.G. and S.G.-G.; Data Analysis, J.G.G. and S.G.-G.; Writing (Review and Editing), S.G.-G. All authors have read and agreed to the published version of the manuscript.

Funding: This research received no external funding.

Conflicts of Interest: The authors declare no conflict of interest.

References

1. Hicks, C.; Mcgovern, T.O.M.; Earl, C.F. A Typology of UK Engineer-To-Order Companies. *Int. J. Logist.* **2001**, *4*, 43–56. [CrossRef]
2. Lacroix, Y.; Major Overcapacity in the Global Steel Industry. Major Overcapacity in the Global Steel Industry. Euler Hermes Economic Research. Economic Insight. 2013. Available online: http://www.eulerhermes.com/mediacenter/Lists/mediacenterdocuments/20131014_economic_insight_steel.pdf (accessed on 17 March 2014).
3. Price, A.H.; Weld, C.B.; El-Sabaawi, L.; Teslik, A.M. Unsustainable: Government Intervention and Overcapacity in the Global Steel Industry. 2016, pp. 1–25. Available online: https://www.wileyrein.com/media/publication/204_Unsustainable-Government-Intervention-and-Overcapacity-in-the-Global-Steel-Industry-April-2016.pdf (accessed on 29 December 2019).
4. Cao, W. An Empirical Study of the Impact of the "Cutting Overcapacity" Policy on the Performance of Steel Industry. *DEStech Trans. Eng. Technol. Res.* **2018**. [CrossRef]
5. Steelmaking Capacity. ECD Steelmaking Capacity Database 2018. Available online: Retrievedfrom:https://stats.oecd.org/Index.aspx?DataSetCode=STI_STEEL_MAKINGCAPACITY (accessed on 29 December 2019).
6. World Steel Association. Steel Statistical Yearbook 2018. Available online: https://www.worldsteel.org/en/dam/jcr:e5a8eda5-4b46-4892-856b-00908b5ab492/SSY_2018.pdf (accessed on 29 December 2019).
7. Hicks, C.; McGovern, T.; Earl, C.F. Supply Chain Management: A Strategic Issue in Engineer to Order Manufacturing. *Int. J. Prod. Econ.* **2000**, *65*, 179–190. [CrossRef]
8. Lee, K.; Ki, J.H. Rise of Latecomers and Catch-Up Cycles in the World Steel Industry. *Res. Policy* **2017**, *46*, 365–375. [CrossRef]
9. Nozomu, K. *Where Is the Excess Capacity in the World Iron and Steel Industry?—A Focus on East Asia and China*; Research Institute of Economy, Trade and Industry: Tokyo, Japan, 2017; No. 17026.
10. Campuzano, F.; Mula, J. *Supply Chain Simulation: A System Dynamics Approach for Improving Performance*; Springer Science & Business Media: Berlin, Germany, 2011.
11. Borshchev, A. Simulation Modeling with Anylogic: Agent Based. *Discret. Event Syst. Dyn. Methods* **2013**, 225–269.
12. Gosling, J.; Naim, M.M. Engineer-To-Order Supply Chain Management: A Literature Review and Research Agenda. *Int. J. Prod. Econ.* **2009**, *122*, 741–754. [CrossRef]
13. Westkämper, E. *Einführung in die Organisation der Produktion*; Springer: Berlin, Germany, 2006.
14. Hinckeldeyn, J.; Dekkers, R.; Altfeld, N.; Kreutzfeldt, J. Bottleneck-Based Synchronisation of Engineering and Manufacturing. In *International Association for Management of Technology IAMOT 2010, Proceedings of the 19th International Conference on Management of Technology, Cairo, Egypt, 8–11 March 2010*; Industry Committee of the International Association for Management of Technology: Cairo, Egypt, 2010.
15. Handfield, R.B. Effects of Concurrent Engineering on Make-To-Order Products. *IEEE Trans. Eng. Manag.* **1994**, *41*, 384–393. [CrossRef]
16. Lengnick-Hall, C.A. Innovation and Competitive Advantage: What We Know and What We Need to Learn. *J. Manag.* **1992**, *18*, 399–429. [CrossRef]
17. Kashan, A.J.; Mohannak, K.; Perano, M.; Casali, G.L. A Discovery of Multiple Levels of Open Innovation in Understanding the Economic Sustainability. A Case Study in the Manufacturing Industry. *Sustainability* **2018**, *10*, 4652. [CrossRef]
18. Rupo, D.; Perano, M.; Centorrino, G.; Vargas-Sanchez, A. A Framework Based on Sustainability, Open Innovation, and Value Cocreation Paradigms—A Case in an Italian Maritime Cluster. *Sustainability* **2018**, *10*, 729. [CrossRef]
19. Baláž, P.; Bayer, J. Impact of China on Competitiveness of EU Steel Industry (Slovakia). In Proceedings of the 4th International Conferenceon European Integration, Ostrava, Czech Republic, 17–18 May 2018; p. 118.
20. Hoitsch, H. *Produktionswirtschaft: Grundlagen einer industriellen betriebswirtschaftslehre*; Vahlen: Munich, Germany, 1985.
21. Schuh, G.; Brosze, T.; Brandenburg, U.; Cuber, S.; Schenk, M.; Quick, J.; Hering, N. Grundlagen Der Produktionsplanung Und-Steuerung. *Prod. Und-Steuer.* **2012**, *1*, 11–293.
22. Nyhuis, P. *Produktionskennlinien—Grundlagen und Anwendungsmöglichkeiten*; Springer: Berlin, Germany, 2008.
23. Wirtschaftslexikon, G. *Stichwort: Produktionsprozessplanung*; Verlag: Wiesbaden, Germany, 2013.

24. Chopra, S.; Meindl, P. Supply Chain Management. Strategy, Planning & Operation. *Das Summa Summ. Des Manag.* **2007**, 265–275. [CrossRef]

25. Schuh, G.; Weber, H.; Kajüter, P. *Logistikmanagement*; Springer: Berlin, Germany, 2013.

26. Zäpfel, G. *Grundzüge Des Produktions-Und Logistikmanagement Walter De Gruyter*; De Gruyter Oldenbourg: Munich, Germany, 2001.

27. Hung, K.T.; Liou, E.D.; Wen, C.P.; Tsai, H.F.; Shi, C.S.; Chang, Y.C.; Lee, Y.J. Decision Support System for Inventory Management by TOC Demand-Pull Approach. In Proceedings of the 2010 IEEE/SEMI Advanced Semiconductor Manufacturing Conference (ASMC), San Francisco, CA, USA, 11–13 July 2010; pp. 23–26.

28. Jodlbauer, H.; Huber, A. Service-Level Performance of MRP, Kanban, CONWIP and DBR Due to Parameter Stability and Environmental Robustness. *Int. J. Prod. Res.* **2008**, *46*, 2179–2195. [CrossRef]

29. Bleicher, K. *Das Konzept Integriertes Management: Visionen-Missionen-Programme*; Campus Verlag: Frankfurt/Main, Germany, 2004.

applied
sciences

Article

Experimental and Numerical Simulation Investigation on Deep Drawing Process of Inconel 718 with and without Intermediate Annealing Thermal Treatments

Unai Ulibarri, Lander Galdos, Eneko Sáenz de Argandoña and Joseba Mendiguren *

Department of Mechanical and Industrial Production, Mondragon Unibertsitatea, Loramendi 4, 20500 Mondragon, Spain; unai.ulibarri@gmail.com (U.U.); lgaldos@mondragon.edu (L.G.); esaenzdeargan@mondragon.edu (E.S.d.A.)
* Correspondence: jmendiguren@mondragon.edu

Received: 30 October 2019; Accepted: 8 January 2020; Published: 13 January 2020

Abstract: The aeronautical industry is moving from high-capacity large-airplane construction to low-capacity small-airplane construction. With the change in the production volume, there is a need for more efficient manufacturing processes, such as stamping/deep drawing. However, the streamlined shape and exotic materials of airplanes pose a challenge to accurate numerical simulation of the manufacturing processes. In the case of the Inconel 718 material, researchers previously proposed numerical models; however, these models failed to take account of some key parameters, such as the degradation of the elastic modulus and intermediate annealing thermal processes. The aim of the present study was to characterize the Inconel 718 material, with and without intermediate annealing thermal treatment (TT) and to propose a suitable model. To evaluate the accuracy of the proposed model, a U-drawing benchmark test was used.

Keywords: Inconel 718; Inco718; springback; bauschinger; autoform

1. Introduction

The aeronautical industry has long been one of the leading players in the development and implementation of new high-added-value, advanced materials [1]. The big margin on manufacturing cost compared to the material and performance has made difficultly shaped components and the extensive use of trial and error methodologies in the manufacturing process characteristic of this sector. Clear examples to reduce that effort are the work of Groche and Backer [2] on stretch forming and the work of Zhan et al. [3] on Ti-alloy tube bending. Global mobility trends, together with tightening of the economy, have led to a departure in aircraft manufacturing from big planes (e.g., A380 and Boeing 747) to the current trend of a large number of smaller aircraft (e.g., A320 or smaller) [4]. This change in product volume has increased the importance of manufacturing costs relative to overall costs within the sector.

Driven by such change, numerous aeronautical component producers have been exploring the use of automotive-based technologies to reduce manufacturing costs, while maintaining quality [5]. In this regard, stamping (or deep drawing) is one of the most cost-effective manufacturing processes for sheet metal components used in aircraft fuselage and engines [6]. The migration from traditional forming processes to mass production stamping base processes has given rise to various issues due to the high springback. Traditionally, aeronautical components are characterized by a very streamlined and small curvature design and are manufactured with high- or very-high-strength material. These two features increase the springback after forming exponentially [7]. To address these issues, thermal treatments (TTs) are commonly employed in an effort to reduce residual stresses or improve formability. The sector is in need of high-volume manufacturing processes that allow for intermediate TTs.

Previous research demonstrated the importance of finite element methods in efficient design and optimization of cold stamping processes [7]. In recent years, much effort has focused on the modelling of aeronautical-grade stamping operations [8]. However, due to the innate secrecy of the sector, little published information on this topic is available. In the case of Inconel 718, some information is available on the elasto–plastic modelling of stamping operations. For example, Algarni et al. [9] studied the ductile fracture of bulk Inconel 718, and Gustafsson et al. [10] modelled the behaviour of the material at intermediate temperatures using kinematic hardening models. The most comprehensive work on modelling Inconel 718 material was published by Ragai et al. [11]. They analysed and modelled the springback behaviour of Inconel 718 used in the aerospace industry. In their hardening model, they assumed constant elastic behaviour, with a Hill48 yield function and a Bauschinger effect. Later research proved that a decrease in the apparent elastic modulus had a critical impact on springback simulations [12]. Other studies showed that this phenomenon, which appeared to be related to dislocation pile-up [13], increased springback after forming [14]. The friction conditions during draw-in of the material were shown to be equally important [7].

To our knowledge, there are no studies on the impact of intermediate annealing TTs on material behaviour. Therefore, in terms of aeronautical requirements, it is not possible to accurately predict the forming behaviour and subsequent springback of components of Inconel 718 after intermediate annealing TTs. In the present study, the behaviour of Inconel 718 material in its as-received (AR) state and post-stretching TT (PSTT) state was analysed and modelled. First, potential microstructural changes due to the annealing TT were analysed. Second, the elasto–plasticity behaviour of the material (i.e., elastic behaviour, plastic yielding, hardening and necking limit) was characterized under both AR and PSTT conditions. Third, the material behaviour was modelled and used as an input for deep drawing simulations. Finally, U-drawing benchmark experimental tests were conducted, and the accuracy of the different models developed was evaluated.

2. Materials and TTs

Inconel 718 material (HAYNES®718; Haynes International) (hereafter, Inco718) was used. The material (1.645 mm thick) was supplied in a solution annealed state. Its chemical composition is shown in Table 1.

Table 1. Chemical composition of the Inco718 material (wt.%).

Al	B	C	Co	Cr	Cu	Fe	Mn	Mo
0.65	0.004	0.049	0.37	18.10	0.04	18.6	0.22	3.07
Ni	P	S	Si	Ti	Ta	Bi	Pb	Ag
52.9	<0.005	<0.002	0.08	1.04	<0.05	<0.00003	<0.0005	<0.0002

The material specifications defined by the material supplier are defined in Table 2.

Table 2. Material specifications provided by the material supplier.

Elastic Modulus (GPa)	σ_y (MPa)	σ_u (MPa)	Elongation (%)	Hardness (HRB)	ASTM
200	419	871	46	92.4	6-8

According to the supplier, Inco718 has a Young's modulus of 200 GPa, with a yielding stress value of 419 MPa and ultimate stress value (real stress at σ_u) of 871 MPa. The material has a ductility (elongation) of 46% at the point of necking, a grain size of 6–8 ASTM and hardness of 92.4 HRB.

Figure 1 shows the difference in the material in its AR state and PSTT state (i.e., after deformation and annealing) in a simple tensile test.

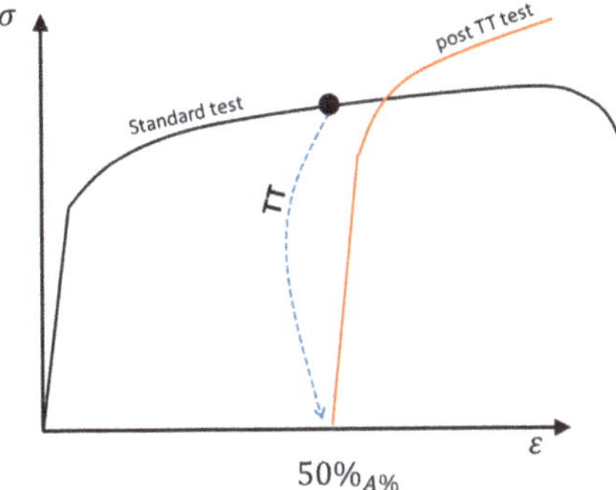

Figure 1. Difference between the as-received (AR) state and post-stretching thermal treatment (PSTT) state shown as a post-TT test.

The standard test corresponds to the AR condition, whereas the PSTT condition corresponds to the material after it had been stretched by up to 50% of its total elongation capability (50% of Rm elongation) and subsequently annealed. In the intermediate annealing process, the holding time was 6 min under 980 °C, followed by air-cooling at room temperature. As the material had an elongation to Rm of about 46% of plastic strain, a limit of 22% of plastic strain was taken as the pre-stretch for the PSTT samples.

To evaluate whether the intermediate annealing process changed the microstructure beyond the relaxation of the dislocations, a microstructural analysis of the material under three different conditions was conducted. Figure 2 shows representative SEM-ETD images of the samples under the various conditions.

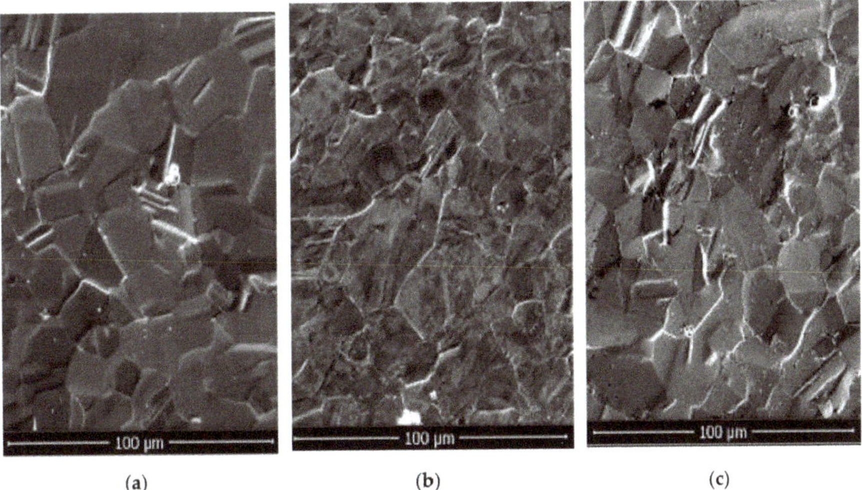

Figure 2. SEM-ETD images of the microstructure of the material under three different conditions. (a) AR condition, (b) after stretching the material to 50% of its stretching capacity and (c) after the intermediate annealing process (PSTT condition). The images were taken in the RD&ND direction (on the section plane between the rolling direction and the normal direction).

The averaged grain size of the samples remains on the 5 ASTM with a deviation of 1 ASTM. On the one hand, this is on the boundary of the supplier specifications, and on the other hand, this denotes that neither the TT and the pre-strain has deformed excessively the grain size.

Based on these observations, there were no obvious microstructural changes in the samples under the different conditions, suggesting that no critical microstructural changes were introduced during the annealing process. The results also indicated that only the dislocation density played a role in the elasto–plastic behaviour of the material. These results were expected as the holding time and temperature specifications used in this study were taken from an aeronautical handbook. Thus, these variables were not expected to modify the microstructure beyond stress relaxation.

Next, we investigated the changes in the behaviour in the elasto–plastic strain–stress state.

3. Characterization of Mechanical Properties of the Material

To characterize the elasto–plastic behaviour of the Inco718 material, we analysed its elastic behaviour and plasticity-induced evolution in this behaviour, followed by an analysis of plastic yielding and plastic flow. The hardening behaviour and Bauschinger effect were then experimentally observed. The necking limit of the material, characterized by its forming limit curve (FLC), was then evaluated. Furthermore, the surface texture and the impact of this texture on the friction coefficient were evaluated.

3.1. Elastic Behaviour

Dislocation pile-up can lead to non-linear hysteresis behaviour during unloading and loading cycles after plastic stretching of material [13]. Figure 3 shows a schematic of the behaviour of Inco718 in a loading–unloading test designed to characterize the material's non-linear behaviour.

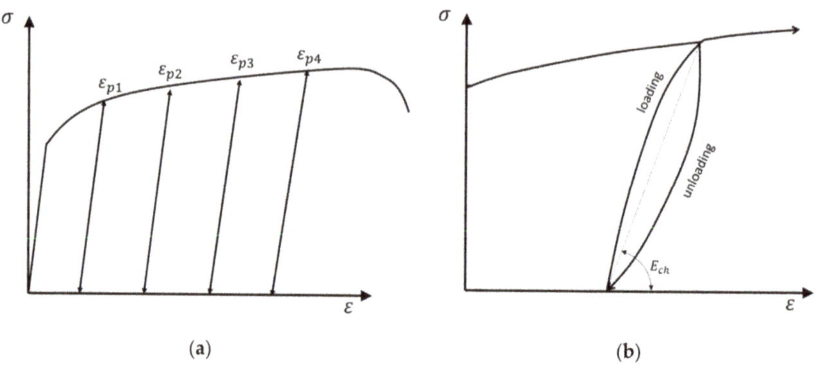

Figure 3. Schematic representation of the elastic modulus degradation testing in where: (**a**) loading–unloading test and (**b**) a hysteresis loop observed during the elastic unloading–loading step.

During this test, a standard tensile sample was stretched until a pre-defined pre-strain value (ε_{p1}). The sample was unloaded until zero stress was reached and then loaded again up to the pre-strain ε_{p1} value. Next, the sample was unloaded again, and the cycle was repeated, increasing the pre-strain level at each loop. A schematic representation of this process and the results obtained are shown in Figure 3a. As shown by an analysis of the elastic unloading–loading (Figure 3b), the unloading (and loading) path followed non-linear behaviour, creating a hysteresis loop. The modulus of the slope linking the start and finishing points of the loop was then measured (i.e., the chord modulus, E_{ch}).

Usually, the value of the chord modulus decreases with an increase in the pre-strain of the loop. In the present study case, the value of the chord modulus in both sets of samples (AR and PSTT) decreased from about 176 GPa at the first loading until around 150 GPa after 7% to 10% of plastic strain. At least three samples were tested under each condition, and the average value is shown in Table 3. The strain measurement of the samples was conducted using high-elongation strain gauges glued to the centre of the tensile specimen.

Table 3. Chord modulus evolution of the Inco718 material.

Pre-Strain	0%	1%	1.5%	2%	3%	4.5%	7%	9%	10%
AR (GPa)	176	160	156	155	152	149	149	149	149
PSTT (GPa)	177	157	155	153	151	148	148	-	-

Figure 4 shows the evolution of the chord modulus (with the deviation of the values as error bars) for both states (AR and PSTT).

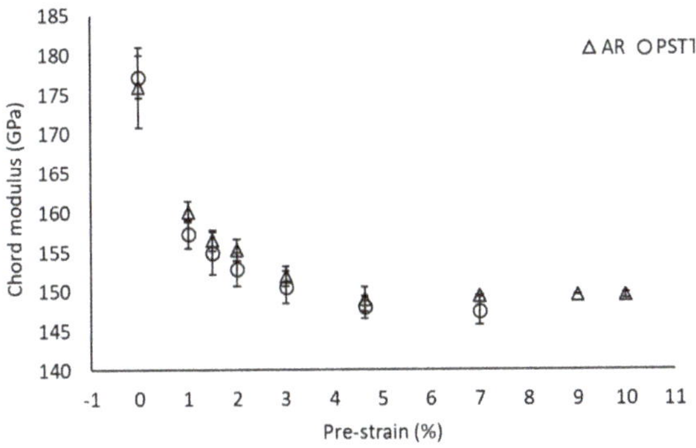

Figure 4. Chord modulus evolution in the AR material and PSTT material.

As shown in Figure 4, similar changes were observed in the chord modulus behaviour of the AR and PSTT samples. The latter may suggest that the annealing process completely dissolved dislocation and that dislocation pile-ups were responsible for the hysteresis loops in the initial material state, at least in terms of the material's elastic behaviour. However, the chord modulus, which is commonly used to denote the equivalent elastic modulus, rapidly decreased in the first 4% to 5% of pre-strain and reached an asymptotic value of around 150 GPa. It is worth noting that the initial equivalent elastic modulus (chord modulus of the first loading) was 20 GPa lower than that specified by the material supplier. This is consistent with previous observations on high-strength steel when measuring using strain gauges and taking into account the non-linearity of the first loading [14,15]. This difference could have an impact on springback simulations [16].

3.2. Plastic Yielding and Plastic Flow

To experimentally evaluate the plastic yielding and plastic flow of the material, we conducted a tensile test. The test parameters were as follows: rolling direction (RD), 45 degrees to the RD (45D) and transversal direction (TD) to the RD. To evaluate the width strain and longitudinal strain simultaneously, a digital image correlation GOM-Aramis 5M system was used. The through-thickness strain was calculated based on the width strain and longitudinal strain values and assuming a constant volume, and the anisotropy coefficient or r-value was calculated. Using linear regression, the values were calculated between 10% and 20% of plastic strain. Table 4 shows the mean values of the three samples. These values were considered representative of the plastic flow of the material.

Table 4. Anisotropy coefficients of the Inco718 material (mean value, where the deviation was less than 0.03 in all cases).

State	RD	45D	TD
AR (-)	0.616	1.043	1.085
PSTT (-)	0.766	0.983	0.981

As shown in Table 4, there was no significant difference between the AR and PSTT behaviour of the samples. The largest difference between both states was found at RD (0.15 of difference). However, the difference was not remarkable from a practical point of view.

To evaluate the role of plastic yielding in triggering a stress state, the R_{p02} was calculated in each direction (RD, 45D and TD). In addition, to analyse the potential non-symmetry of the plastic contour, plastic yielding triggering stresses were calculated under compression loads using an anti-buckling device [17]. Table 5 shows the tension and compression values in each direction.

Table 5. Yield stress values of the Inco718 material under different directions and load conditions.

State	RD		45D	TD	
	Tensile	Compression	Tensile	Tensile	Compression
AR (MPa)	431 ± 2.1	441 ± 1.9	448 ± 0.8	449 ± 2.0	457 ±1.9
PSTT (MPa)	422 ± 2.3	435 ± 4.0	434 ± 2.7	434 ± 2.2	456 ± 2.5

Differences under 3% were found between the values of AR and PSTT. In terms of non-symmetry of the plastic contour, the difference was less than 5%.

Similar to the plastic flow and elastic behaviour, no appreciable difference was observed between the AR and PSTT states. Therefore, the use of a symmetric surface will not introduce too much error, as reviewed by Lewandowski et al. [18].

3.3. Hardening Behaviour

Two main characteristics of the hardening behaviour of a material are typically evaluated: stress–strain behaviour under monotonic stretching and the existence of a Bauschinger effect under strain path changes.

Figure 5 shows both characteristics for Inco718. Figure 5a illustrates the characteristic stress–strain relation in samples ($n = 3$) of the stretched material in both states (AR and PSTT), and Figure 5b shows the stress–strain relation under a strain path change at 2% of pre-strain.

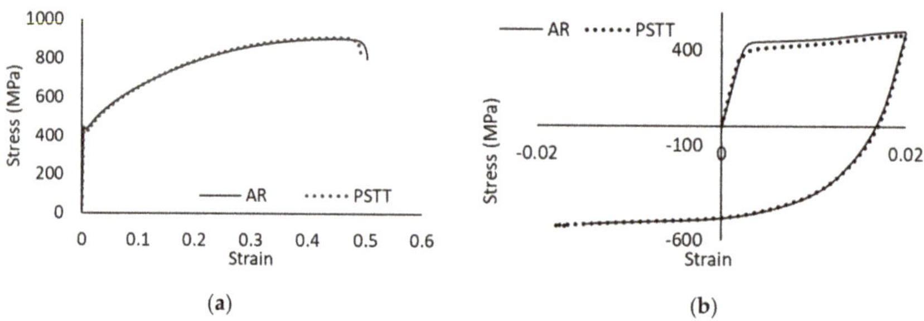

Figure 5. Hardening behaviour of the Inco718 material. (**a**) Monotonic stress–strain relation (engineering stress-strain) and (**b**) strain path change and stress–strain relation.

As shown in Figure 5a,b, both material states (AR and PSTT) exhibited the same hardening behaviour. The similar hardening behaviour is in accordance with the values found in the previous experimental tests. The Bauschinger effect was observed in both states (Figure 5b), with early yielding observed almost when the stress state started in compression. This phenomenon could be critical for accurate predictions of stresses under strain path changes.

3.4. Formability

The necking limit, often denoted by the FLC (forming limit curve), is a critical characteristic of all sheet material subjected to deep drawing operations. The FLC represents the limiting stretching value where necking occurs under different strain paths (major and minor strain ratios). The FLC is determined in a Nakajima test, according to the ISO 12004 standard. The aim of intermediate annealing processes is to restore the initial material state and, therefore, improve the stretchability of the material allowing a restoration of the formability diagram.

Formability was analysed in Nakajima experiments, following the ISO 12004 standard for testing and post-processing methodology. Figure 6a shows the geometries of the samples, with each of these leading to a different strain path.

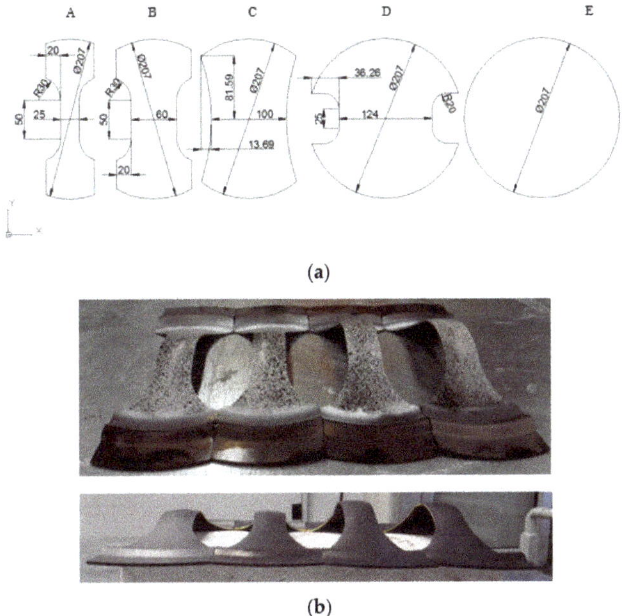

Figure 6. Forming limit of Inco718 samples. (**a**) Sample geometry and (**b**) pre-strain samples.

The testing procedure and sample preparation for the PSTT samples in the present study differed somewhat to those used in previous studies. The procedure was as follows: First, the standard FLC characterization of the AR samples was conducted. Following this procedure, a necking limiting point was obtained for each sample geometry (A, B, C, D and E). To conduct the pre-strain of the PSTT samples, the samples with different geometries were then drawn until 25% of the strain path of that geometry was reached. Next, the intermediate annealing process was carried out, and the samples were subsequently drawn until necking only, taking into account the strain values introduced in the second drawing step. For example, if a sample that followed a 'plane strain' path had a limiting strain at 0.3 of major strain (in the AR condition), for pre-stretching, the sample was drawn until 0.075 of major strain was obtained. The sample was then annealed and subsequently drawn until its necking limit was reached. If the sample reached 0.3 of the major strain after annealing, 0.075 was its pre-strain major strain, and an extra 0.3 of major strain was withstood before necking occurred. Thus, the total deformation was 0.375 using the annealing procedure. Figure 6b shows the samples after pre-strain.

To evaluate the capability of stretching of the Inco718 material under both AR and PSTT conditions, the FLC value of the AR material and that of the PSTT, taking into account only the strain introduced after intermediate annealing, were calculated. The results are shown in Figure 7.

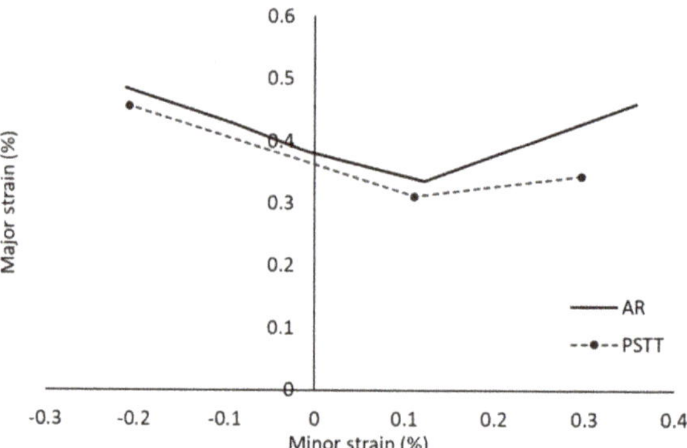

Figure 7. FLC of the Inco718 material. The continuous lines show the FLC of the AR necking limit, and the dashed lines denote the PSTT necking limit.

In accordance with standard industrial procedures for the cold stamping of materials, a graphite-based lubricant was used and a centred crack on the middle of the dome. GOM ARAMIS software was used for the online strain measurement and ISO standard application.

As shown in Figure 7, the minimum value point, theoretically the 'plane strain' point, was on the right side of the graph, even with a material thickness of 1.645 mm and a punch of 100 mm in diameter. These values may be explained by the theory of Min et al. [19]. As the objective of the present study was to compare the behaviour of both states (AR and PSTT), these values were accepted and included in the analysis.

As shown by the left part of the FLC, in contrast to the findings of previous experiments, there is a difference in the necking capacity on the biaxial strain path. Li et al. [20] observed a similar trend in a study on AA5182-O before and after intermediate annealing. They argued that the necking limiting strain under the biaxial strain state was close to the fracture point and that this made it difficult to determine whether the evaluated value was the necking point. With the results in hand, more comprehensive analyses are needed to shed light on this issue.

3.5. Friction Characterization

Apart from material behaviour, friction between the die and sheet also plays a key aspect in drawing operations. According to previous research, this friction was highly dependent on the lubricant, sheet, die surface roughness and material, as well as on contact pressure, sliding velocity and temperature [7].

In this study, the die material used was F-522T. The dies were polished in a toolmaker partner to give a final surface quality of 0.22 μm Ra and 1.34 μm Rz. In a strip drawing test, the dies were used in conjunction with strips of the Inco718 material under different pressures ranging from 1 MPa to 40 MPa at a constant speed of 10 mm/s at room temperature.

Prior to the strip drawing test, surface strips of each sample covering an area of 3250×3040 μm^2 was analysed using a Sensofar S Neos with a 20× objective, sampling of 0.65 μm and vertical resolution of 8 nm. Figure 8 shows the topographical analysis of the surfaces of the samples. Figure 8a,c shows the values for the AR sample, and Figure 8b,d depicts the values for the PSTT sample.

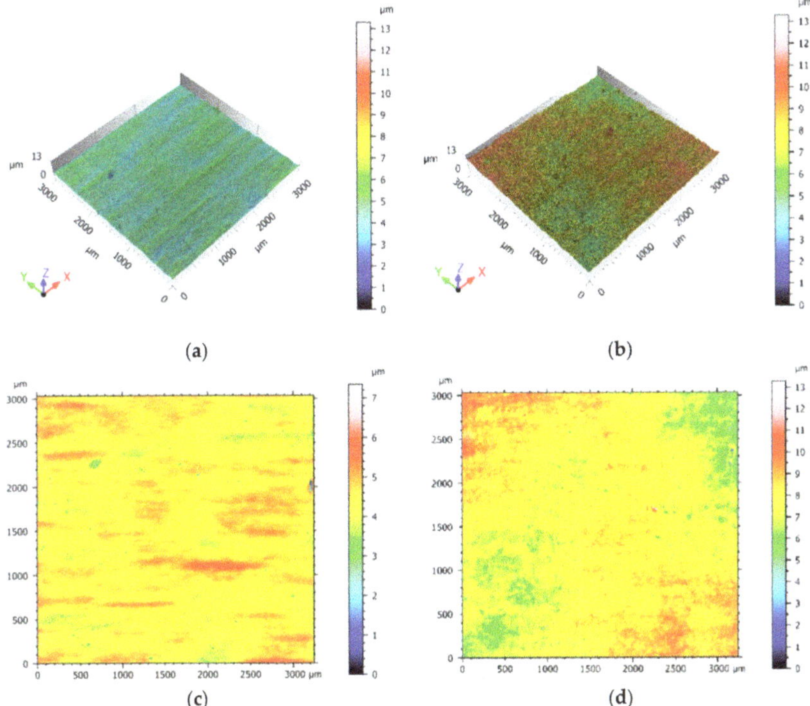

Figure 8. Topographic view of the material surface. (**a,c**) 3D and 2D representations of the surface of the AR material, respectively. (**b,d**) 3D and 2D representations of the surface of the PSTT material, respectively.

To evaluate differences between the surfaces, the Sq (the root mean square value of ordinate values) and Sdr (the developed interfacial area ratio) values of the surfaces were calculated. The results are shown in Table 6.

Table 6. Quantitative comparison of the surface of the AR material and that of the PSTT material after 7% pre-strain.

State	Sq (μm)	Sdr (μm)
AR	0.38 ± 0.04	0.48 ± 0.02
PSTT	0.79 ± 0.02	1.01 ± 0.06

Pre-strain of the sample and subsequent TT modified the surface roughness, increasing the Sq value from 0.38 to 0.79 and the Sdr value from 0.48 to 1.01. The increase in these values could affect the friction and wear behaviour of the material.

To evaluate the impact of changes in thee values on friction behaviour, strip drawing tests were conducted. In the strip drawing tests, G-Start graphite spray and Condat Vicafil TFH4002 lubricant, which is commonly used in material drawing operations, were used. During the test, a strip of the sheet material under analysis is pressed between two square flat dies in order to apply the desired contact pressure. Once the pressure is achieved, the strip is pulled, and a relative displacement between the dies and the sheet is generated in which both forces—the puling force and the normal force—are measured in order to calculate the existing friction coefficient on the tribological system. Further details of the system can be found in [21]. The friction coefficient values obtained in the strip drawing tests are shown in Table 7.

Table 7. Strip drawing test friction coefficient values for both material surfaces at different contact pressures.

State	1 MPa	2.5 MPa	5 MPa	10 MPa	15 MPa	20 MPa	25 MPa	30 MPa	40 MPa
AR	0.037	0.022	0.009	0.007	-	0.011	-	0.014	0.016
PSTT	0.034	0.020	0.013	0.010	0.011	-	0.013	-	-

As shown in Table 7, even with an increase in surface roughness, the differences in the friction coefficient values of both material states (AR and PSTT) were negligible. Furthermore, the friction coefficient value obtained for the combination of materials and lubricant was low (<0.08). Changes in the coefficient of friction are shown in Figure 9.

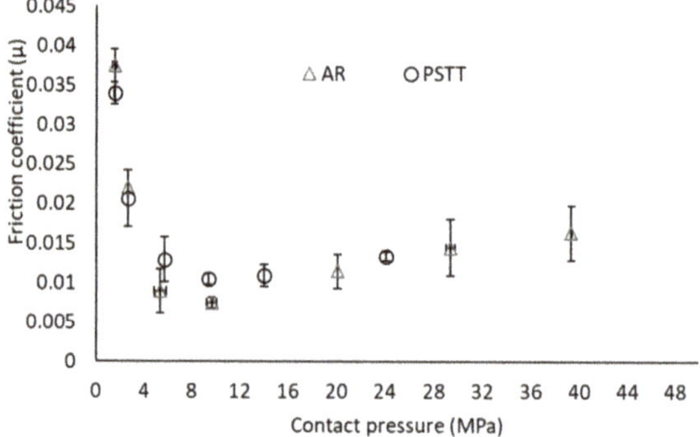

Figure 9. Evolution of the coefficient of friction measured during the strip drawing test under different contact pressures.

As shown in Figure 9, the friction coefficient rapidly decreased in the first 8 MPa, followed by saturation at a value of around 0.01. In the figure, the slight increase starting at 20 MPa may have been due to lubricant slippage. The high viscosity of the used lubricant usually creates a lubricant film, preventing contact between both surfaces. Therefore, in the first range of pressures (from 0 MPa to 20 MPa) the interaction behaviour is related to the behaviour of the lubricant under different pressures. However, under high pressures, the lubricant slips between the surfaces, remaining only in the valleys of the topography, and the tribological system changes as the contact of asperities can develop. However, further studies are necessary in order to clarify the origin of the increase in friction behaviour.

Based on the results obtained, with the exception of the necking limit for the biaxial strain path, in all other aspects, the intermediate annealing process restored the initial material state. Therefore, the behaviour of the PSTT material was similar to that of the AR material.

4. Material Modelling

After characterizing the properties of the material, deep drawing finite element software AutoForm®, a commonly used commercial software, was used. The work focuses on models available in the software (with which we will later simulate the benchmark test). Similar to the material characterization, the modelling covered the elastic behaviour, plastic yielding and flow and hardening of the material.

4.1. Elastic Behaviour Modelling

There are three levels of simulation complexity when modelling the elastic behaviour of metals in deep drawing. The simplest type of simulation complexity involves the use of a constant modulus (Young's modulus) with linear elastic behaviour. The second level of simulation complexity involves the use of linear elastic behaviour but with an evolving elastic modulus. Usually, the chord modulus is taken as a reference [22]. The third level of simulation complexity involves the use of non-linear elastic behaviour with evolution due to plasticity [15]. In terms of the last level, there are three main models: the models of Yoshida [23], Wagoner [24] and Mendiguren [14]. However, none of these models are currently cost-efficient for use in industrial deep drawing simulations. The main model used in this second level of complexity is the model known as Yoshida's model. This model can be found in many scientific contributions in the last decade (e.g., [22]). The model predicts an asymptotic decrease in the elastic modulus defined as below:

$$E_{ch} = E_0 - (E_0 - E_a)(1 - \exp(-\gamma \bar{\varepsilon}^{P})), \tag{1}$$

where E_{ch} represents the chord modulus, E_0 is the initial modulus, E_a is the saturated modulus, $\bar{\varepsilon}^{P}$ is the accumulated plastic strain, and γ is the saturation rate parameter of the model.

In the present study, the model coefficients that best suited the behaviour described in Figure 4 were as follows: E_0 of 177 GPa, E_a of 150 GPa and γ of 100.

4.2. Yield Criteria

Different hypotheses can be assumed regarding the flow rule, associated flow role or non-associated flow rule [25]. Although some researchers favour the non-associated assumption, the association of the plastic potential to the yield criteria is commonly assumed. The yield criteria not only have to satisfy the limit stress states where plasticity is triggered but also have to govern the plastic flow of the material due to the normality rule.

Numerous yield functions, such as Hill49, Yld89, Yld2000-2D and BBC2008, have been formulated in the last decade [26]. In the present study, as proposed by previous authors [2], the Hill48 function was used to model the yielding behaviour of Inco718, as below:

$$2(\sigma_e)^2 = (G+H)\sigma_{11}^2 + (F+H)\sigma_{22}^2 - 2H\sigma_{11}\sigma_{22} + 2N\sigma_{12}^2, \tag{2}$$

where G, H, F and N are the model coefficients. At stress level, σ_{ij} correspond to the stress tensor ij component and σ_e represents the equivalent stress. When defining the coefficient to fit the r-values of the material, they were calculated as below:

$$G + H = 1, \tag{3}$$

$$F = (r_0)/(r_{90}(1 + r_0)), \tag{4}$$

$$G = 1/(r_0 + 1), \tag{5}$$

$$N = ((1 + 2r_{45})(r_0 + r_{45}))/(2r_{90}(1 + r_0)). \tag{6}$$

In the above equations, r_0, r_{90} and r_{45} represent the anisotropy coefficient at RD, TD and 45D, respectively.

Once the model was fitted to represent the r values, it was checked to determine whether it accurately represented normalized yield stress ratios. In Figure 10, the accuracy of the Hill48 prediction is shown.

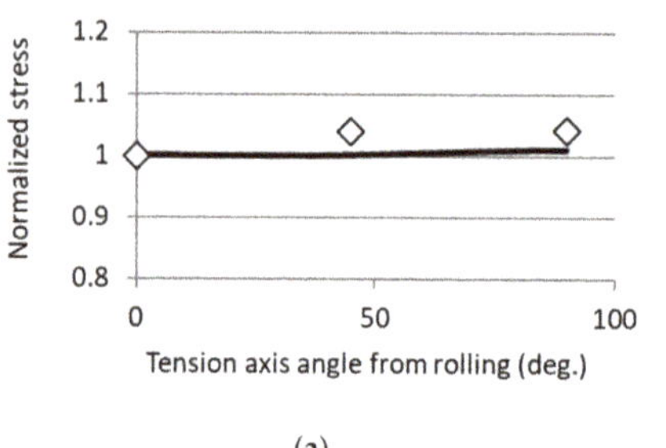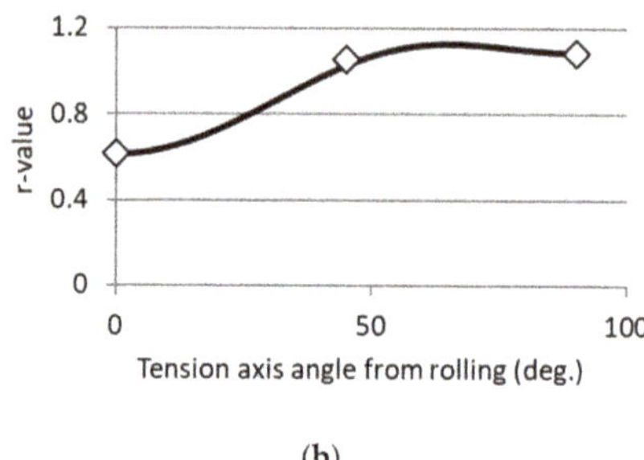

(a)   (b)

Figure 10. Anisotropic behaviour of the material in where: (**a**) normalized yield stresses using the Hill48 yielding function and (**b**) *r*-values. The markers show the experimental values, and the continuous line shows the model prediction.

Although the model was fitted using only the *r*-values, it predicted the yield stress relatively well. The low level of anisotropy at the yield stress values suggested that the model was representative of the behaviour of Inco718.

4.3. Hardening Law

There are various hardening laws described in the literature (e.g., Voce, Swift, Ludwik, Hollomond, Ghosh and Hockett-Sherby) [27]. Ragai [11] proposed the use of the hardening models of Ludwik and Hollomond. In the present study, after analysing the fitting accuracy of each model to the specific behaviour of Inco718 [11], we selected the model of Swift. According to this model, the hardening stress, σ_y, was defined as follows:

$$\sigma_y = K(\varepsilon_0 + \bar{\varepsilon}^p)^n, \tag{7}$$

where K, ε_0 and n are the model parameters, with values of 1.951 MPa, -0.09 and 0.299, respectively. The idea is to use the experimental data up to the necking limit and to use Swift's model to extend that data up to the necessary strain. In this way, the negative value of ε_0 did not pose a problem, as the model only started from around 40% of plastic deformation.

Previous studies used different models to analyse the Bauschinger effect. As the present study used AutoForm software, we used the kinematic hardening model included with the software. Using the fitting tool of the software, the model parameters were 0.005 for the κ parameter and 0.29 for the ξ parameter.

5. Benchmark Analysis

To validate the developed models; a benchmark test was carried out under different holding forces and material states (AR and PSTT). The accuracy of the predicted final shapes (i.e., after springback) with different material models was then evaluated.

5.1. U-Drawing Test Set-Up

The U-drawing test is a commonly used benchmark test for springback simulations. In this test, a 90-degree square U channel is first drawn. When the material is released from the die, characteristic springback occurs [28]. The results of the test are primarily controlled by the die geometry, stroke and blank holding force (BHF). Figure 11 shows a schematic representation of the U-drawing test set-up used in this study.

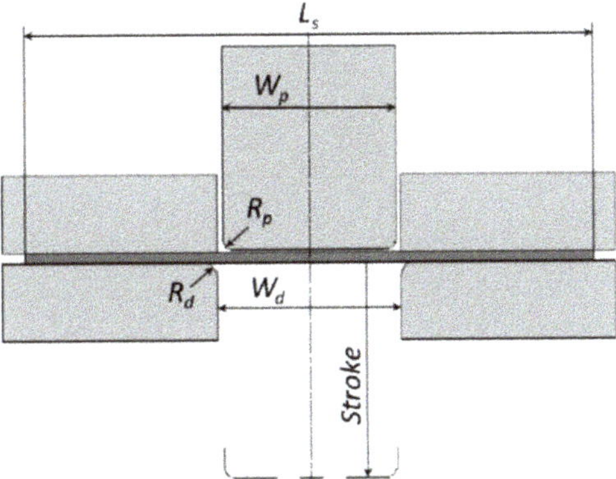

Figure 11. U-drawing test schematic.

During the test, a strip 330 mm long and 110 mm wide was clamped between the blank holder and the die (radius of 8 mm and opening of 106 mm) and then drawn with the punch (radius of 5 mm and width of 100 mm). Details of the test set-up are shown in Table 8.

Table 8. Parameters of the U-drawing experimental test.

W_p (mm)	R_p (mm)	R_d (mm)	W_d (mm)	L_s (mm)	Stroke (mm)
100	5	8	106	330	35-70

Two different BHFs were used to try to generate different strain states during the drawing step: a low BHF (L-BHF) of 8 t and a high BHF (H-BHF) of 28 t.

In the benchmark test, the pre-strain conditions were in accordance with industrial standards. In the AR test condition, the sheet was directly drawn to 70 mm of stroke. In contrast, in the PSTT condition, the sheet was first drawn to 35 mm of stroke. The resulting channel from that step was then annealed following intermediate annealing conditions. Finally, the annealed component was drawn up to 70 mm.

The resulting channels were measured using a coordinate-measuring machine to obtain the final post-springback profile. In this step, the channels were clamped in a central line (a symmetry line), and the profile was then measured.

Figure 12 shows the resultant profiles of the AR and PSTT materials under the different BHFs (L-BHF and H-BHF). Only a representative profile of three tested samples is shown.

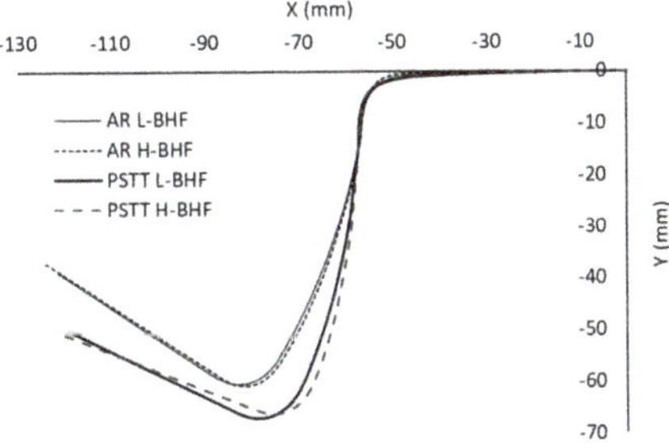

Figure 12. Experimental representation of the resultant channels after springback in the AR and PSTT materials under low blank holding forces (L-BHFs) and high blank holding forces (H-BHFs).

As shown in Figure 12, the holding force did not have a strong impact on springback. Although there was a slight difference in springback of the PSTT material at L-BHFs and H-BHFs, the difference could be due to the positioning on the coordinate-measuring machine scatter. The intermediate TT had a marked impact on the final springback of the U channel.

5.2. Analysed Material Models

Various models of the material and its components (i.e., elasticity, yielding and hardening) were developed and analysed. Table 9 provides a summary of the analysed models. The most basic model (Iso model) included a constant elastic modulus with Hill48 yielding and isotropic hardening. Another model, the kinematic model, included the Bauschinger effect of the material and hardening. A third model was composed of the basic model (Iso) but included a decrease in the elastic modulus (Young). The last and more complex model, referred to as 'full', took into account both the decrease of the elastic modulus and the Bauschinger effect.

Table 9. Summary of the analysed models.

Model Designation	Elasticity	Yielding	Hardening
Iso	Constant	Hill48	Isotropic
Kinematic	Constant	Hill48	Isotropic + kinematic
Young	Variable	Hill48	Isotropic
Full	Variable	Hill48	Isotropic + kinematic

5.3. Friction Model

As shown in Figure 9, the friction coefficient changed with the contact pressure. This type of evolving friction coefficient can be modelled in a variety of ways. On the one hand, one could assume a constant coefficient was valid for a certain range. On the other hand, specific variable friction coefficient laws, such as that of Filsek, could be used [7].

The contact pressure generated by the L-BHF and H-BHF (σ_p) was calculated, taking into account that the surface of the sheet in contact with the die/blank holder declined during the drawing. According to the calculation, the L-BHF pressure varied between 4 MPa and 8.25 MPa, and the H-BHF pressure varied between 12 MPa and 27 MPa. When the friction coefficient values (μ) were analysed at the pressure ranges used in the present study (Figure 9), the coefficient was approximately constant.

$$\text{BHF} = 8\,\text{t} \rightarrow \sigma_p = 4.0 - 8.25\,\text{MPa} \rightarrow \mu = 0.009, \tag{8}$$

$$\text{BHF} = 28\,\text{t} \rightarrow \sigma_p = 12.0 - 27.0\,\text{MPa} \rightarrow \mu = 0.0115. \tag{9}$$

Thus, a constant coefficient of friction was used in the simulation. One coefficient was used in the L-BHF models, and a different one was used in the H-BHF models.

5.4. FEM Model

As previously stated, Autoform® commercial software was used for the numerical simulations. The tools were assumed to be rigid in contrast to the sheet that was discretized in first-order four-node shell elements. These elements were numerically integrated through thickness using 11 integrations points with six levels of refinement. The average size of the elements was defined to be 20 mm. The standard Autoform® 'final validation' convergence tolerances were used, aiming at reproducing the industrial standard.

6. Results and Discussion

In the present study, we compared the springback shape predicted by the models and that experimentally obtained. Figure 13 shows the comparison for the AR H-BHF as a reference. To illustrate the differences between the models in terms of springback predictions, the whole channel is shown in Figure 13a, where only the flange area of the left side of the channel is depicted in Figure 13b. In these figures, X and Y represent space coordinates.

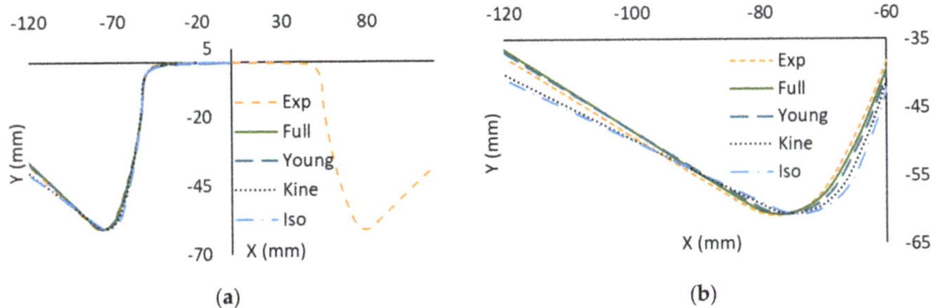

Figure 13. Springback prediction of the different models for the AR H-BHF case. (**a**) Whole profile and (**b**) left side of the profile.

As can be seen from the results, the full and Young models had the highest accuracy. To evaluate the accuracy of each model in every case, the springback angle $\Delta\theta_i$ (final angle minus 90 degrees of the forming angle) was evaluated for each model and analysed case. The angle measurement followed the protocol defined in the Numisheet benchmark of 1993 [28]. Table 10 summarises the values of the springback angles.

Table 10. Summary of the springback angles after forming for each model and experiment.

Material State	Model	L-BHF		H-BHF	
		$\Delta\theta_1$	$\Delta\theta_2$	$\Delta\theta_1$	$\Delta\theta_2$
AR	Exp	7.6	3.45	7.28	3.45
	Full	7.31	2.14	7.02	2.56
	Young	6.92	1.82	6.39	1.12
	Kine	5.44	1.85	4.66	0.57
	Iso	4.84	0.29	4.36	0.34
PSTT	Exp	2.04	4.13	1.56	3.85
	Full	1.18	3.65	0.77	3.02
	Young	1.06	1.15	0.53	1.24
	Kine	0.25	0.98	0.43	0.75
	Iso	0.18	−0.04	0.21	0.67

In the case of the PSTT, the simulations were carried out as follows:

(a) The first drawing was simulated using the AR material model (35 mm drawing).
(b) The springback of that step was then simulated.
(c) Next, the post-springback geometry was imported, without any residual stress or strain, and the second drawing (until reaching a drawing depth of 70 mm) was performed with the AR material model.
(d) Finally, the last springback was simulated.

To evaluate the accuracy of each model, the error of accuracy (in %) between the predicted springback angle and experimental springback angle was calculated. Figures 14 and 15 show the accuracy of the springback angle prediction for $\Delta\theta_1$ and $\Delta\theta_2$, respectively.

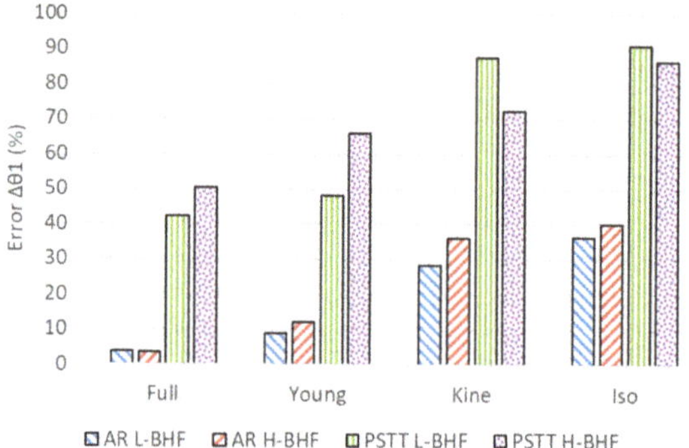

Figure 14. Accuracy of the different models in terms of the Δθ1 springback angle.

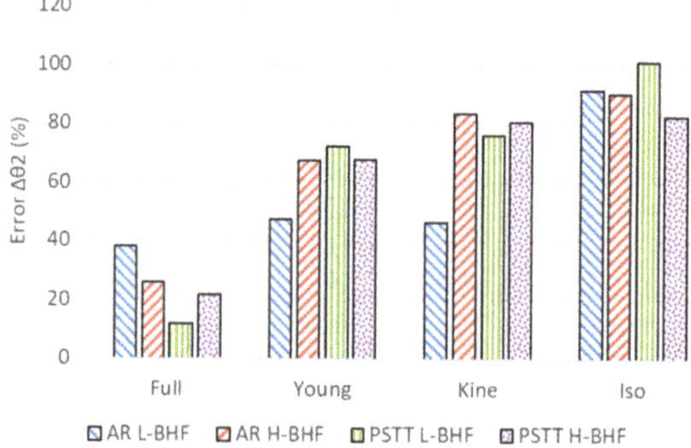

Figure 15. Accuracy of the different models in terms of the Δθ2 springback angle.

From the results, it can be concluded that, in every case, the full model was the most accurate, followed by the Young model. This was expected as the full model was the most complete one. However, for the PSTT, at a H-BHF, errors of up to 50% were found in the Δθ1. In terms of the springback angle, the angles differed in the various models between 91.57 degrees and 90.77 degrees. However, these values were within the accuracy of springback simulations of sheet metal forming [16].

7. Conclusions

In this study, Inco718 material was characterized in a deep drawing process with and without intermediate annealing TTs. In addition, material models were developed, and their accuracy was evaluated using a U-drawing benchmark test. Based on the findings, the following conclusions can be drawn:

- The intermediate annealing process restored the initial properties of the material, with the exception of the biaxial necking limit.
- Failure to restore the biaxial necking behaviour could be due to fracture measurements; further investigations are necessary to clarify this issue.
- The apparent elastic modulus of the material markedly decreased reaching the 150 GPa after 4% to 5% of plastic pre-strain.
- Although the flow behaviour of the material was anisotropic, isotropic-like behaviour was observed under conditions of yield stress.
- The tension compression test revealed an important impact of the Bauschinger effect.
- Taking into account all the features of the material. The evolution of the elastic modulus and the Bauschinger effect was found to be the key phenomena to model.

- Low-friction behaviour was obtained using the tool material, sheet material and lubricant combination.
- Although the roughness of the material increased during the pre-strain process, subsequent TT did not appear to affect the friction coefficient.

The main contribution of this work is to provide the community with a valid model to simulate the deep drawing process of the Inco718 material, with and without intermediate annealing steps.

Author Contributions: Experimental characterization of the material under different TT, U.U.; Numerical simulation of the process using AutoForm software, L.G.; Benchmark experimental testing, E.S.d.A.; Material model development and project supervision, J.M. All authors have read and agreed to the published version of the manuscript.

Funding: This research received no external funding.

Conflicts of Interest: The authors declare no conflicts of interest.

References

1. Bridges, D.; Xu, R.; Hu, A. Microstructure and mechanical properties of Ni nanoparticle-bonded Inconel 718. *Mater. Des.* **2019**, *174*, 107784. [CrossRef]

2. Groche, P.; Bäcker, F. Springback in stringer sheet stretch forming. *CIRP Ann.-Manuf. Technol.* **2013**, *62*, 275–278. [CrossRef]

3. Zhan, M.; Wang, Y.; Yang, H.; Long, H. An analytic model for tube bending springback considering different parameter variations of Ti-alloy tubes. *J. Mater. Process. Technol.* **2016**, *236*, 123–137. [CrossRef]

4. Cansino, J.M.; Román, R. Energy efficiency improvements in air traffic: The case of Airbus A320 in Spain. *Energy Policy* **2017**, *101*, 109–122. [CrossRef]

5. Mertin, C.; Stellmacher, T.; Schmitz, R.; Hirt, G. Enhanced springback prediction for bending of high-strength spring steel using material data from an inverse modelling approach. *Procedia Manuf.* **2019**, *29*, 153–160. [CrossRef]

6. Palm, C.; Vollmer, R.; Aspacher, J.; Gharbi, M. Increasing performance of hot stamping systems. *Procedia Eng.* **2017**, *207*, 765–770. [CrossRef]

7. Sigvant, M.; Pilthammar, J.; Hol, J.; Wiebenga, J.H.; Chezan, T.; Carleer, B.; van den Boogaard, T. Friction in sheet metal forming: Influence of surface roughness and strain rate on sheet metal forming simulation results. *Procedia Manuf.* **2019**, *29*, 512–519. [CrossRef]

8. Sirvin, Q.; Velay, V.; Bonnaire, R.; Penazzi, L. Mechanical behaviour modelling and finite element simulation of simple part of Ti-6Al-4V sheet under hot/warm stamping conditions. *J. Manuf. Process.* **2019**, *38*, 472–482. [CrossRef]

9. Algarni, M.; Bai, Y.; Choi, Y. A study of Inconel 718 dependency on stress triaxiality and Lode angle in plastic deformation and ductile fracture. *Eng. Fract. Mech.* **2015**, *147*, 140–157. [CrossRef]

10. Gustafsson, D.; Moverare, J.J.; Simonsson, K.; Sjöström, S. Modeling of the constitutive behavior of Inconel 718 at intermediate temperatures. *J. Eng. Gas Turbines Power* **2011**, *133*, 094501. [CrossRef]

11. Ragai, I. Experimental and Finite Element Investigation of Springback of Aerospace/Automotive Sheet Metal Products. ProQuest Dissertations and Theses. Ph.D. Thesis, McGill Univeristy, Montreal, QC, Canada, 2006.

12. Aerens, R.; Vorkov, V.; Duflou, J.R. Springback prediction and elasticity modulus variation. *Procedia Manuf.* **2019**, *29*, 185–192. [CrossRef]

13. Kupke, A.; Madej, L.; Hodgson, P.D.; Weiss, M. Experimental in-situ verification of the unloading mechanics of dual phase steels. *Mater. Sci. Eng. A* **2019**, *760*, 134–140. [CrossRef]

14. Mendiguren, J.; Trujillo, J.J.; Cortés, F.; Galdos, L. An extended elastic law to represent non-linear elastic behaviour: Application in computational metal forming. *Int. J. Mech. Sci.* **2013**, *77*, 57–64. [CrossRef]

15. Mendiguren, J.; Cortés, F.; Gómez, X.; Galdos, L. Elastic behaviour characterisation of TRIP 700 steel by means of loading-unloading tests. *Mater. Sci. Eng. A* **2015**, *634*, 147–152. [CrossRef]

16. Wagoner, R.H.; Lim, H.; Lee, M.-G. Advanced Issues in springback. *Int. J. Plast.* **2013**, *45*, 3–20. [CrossRef]

17. Kuwabara, T.; Kumano, Y.; Ziegelheim, J.; Kurosaki, I. Tension-compression asymmetry of phosphor bronze for electronic parts and its effect on bending behavior. *Int. J. Plast.* **2009**. [CrossRef]

18. Lewandowski, J.J.; Wesseling, P.; Prabhu, N.S.; Larose, J.; Lerch, B.A. Strength differential measurements in IN 718: Effects of superimposed pressure. *Metall. Mater. Trans. A* **2003**, *34*, 1736–1739. [CrossRef]

19. Min, J.; Stoughton, T.B.; Carsley, J.E.; Lin, J. Compensation for process-dependent effects in the determination of localized necking limits. *Int. J. Mech. Sci.* **2016**, *117*, 115–134. [CrossRef]

20. Li, J.; Carsley, J.E.; Stoughton, T.B.; Hector, L.G.; Hu, S.J. Forming limit analysis for two-stage forming of 5182-O aluminum sheet with intermediate annealing. *Int. J. Plast.* **2013**, *45*, 21–43. [CrossRef]

21. Gil, I.; Mendiguren, J.; Galdos, L.; Mugarra, E.; Saenz de Argandoña, E. Influence of the pressure dependent coefficient of friction on deep drawing springback predictions. *Tribol. Int.* **2016**, *103*, 266–273. [CrossRef]

22. Eggertsen, P.-A.; Mattiasson, K. On the identification of kinematic hardening material parameters for accurate springback predictions. *Int. J. Mater. Form.* **2011**, *4*, 103–120. [CrossRef]

23. Sumikawa, S.; Ishiwatari, A.; Hiramoto, J.; Yoshida, F.; Clausmeyer, T.; Tekkaya, A.E. Stress state dependency of unloading behavior in high strength steels. *Procedia Eng.* **2017**, *207*, 179–184. [CrossRef]

24. Lee, J.; Lee, J.-Y.; Barlat, F.; Chung, K.; Wagoner, R.H.; Lee, M.-G. Extension of quasi-plastic-elastic approach to incorporate complex plastic flow behavior an application to springback of advanced high-strength steels. *Int. J. Plast.* **2013**, *45*, 140. [CrossRef]

25. Park, N.; Stoughton, T.B.; Yoon, J.W. A criterion for general description of anisotropic hardening considering strength differential effect with non-associated flow rule. *Int. J. Plast.* **2019**. [CrossRef]

26. Mendiguren, J.; Galdos, L.; de Argandoña, E.S. On the plastic flow rule formulation in anisotropic yielding aluminium alloys. *Int. J. Adv. Manuf. Technol.* **2018**, *99*, 255–274. [CrossRef]

27. Agirre, J.; Galdos, L.; Saenz de Argandoña, E.; Mendiguren, J. Hardening prediction of diverse materials using the digital image correlation technique. *Mech. Mater.* **2018**, *124*, 71–79. [CrossRef]

28. Galdos, L.; De Argandoña, E.S.; Mendiguren, J.; Gil, I.; Ulibarri, U.; Mugarra, E. Numerical simulation of U-drawing test of Fortiform 1050 steel using different material models. *Procedia Eng.* **2017**, *207*, 137–142. [CrossRef]

applied sciences

Article

Smart Multi-Sensor Monitoring in Drilling of CFRP/CFRP Composite Material Stacks for Aerospace Assembly Applications

Roberto Teti [1,2,*], Tiziana Segreto [1,2], Alessandra Caggiano [2,3] and Luigi Nele [1]

[1]   Department of Chemical, Materials & Industrial Production Engineering, University of Naples Federico II, Piazzale Tecchio 80, 80125 Naples, Italy; tsegreto@unina.it (T.S.); luigi.nele@unina.it (L.N.)

[2]   Fraunhofer Joint Laboratory of Excellence on Advanced Production Technology (Fh J_LEAPT UniNaples), Piazzale Tecchio 80, 80125 Naples, Italy; alessandra.caggiano@unina.it

[3]   Department of Industrial Engineering, University of Naples Federico II, Piazzale Tecchio 80, 80125 Naples, Italy

*   Correspondence: roberto.teti@unina.it

Received: 17 December 2019; Accepted: 10 January 2020; Published: 21 January 2020

Abstract: Composite material parts are typically laid out in near-net-shape, i.e., very close to the finished product configuration. However, further machining processes are often required to meet dimensional and tolerance requirements. Drilling, edge trimming and slotting are the main cutting processes employed for carbon fiber-reinforced plastic (CFRP) composite materials. In particular, drilling stands out as the most widespread machining process of CFRP composite parts, chiefly in the aerospace industrial sector, due to the extensive use of mechanical joints, such as rivets, rather than welded or bonded joints. However, CFRP drilling is markedly challenging: due to CFRP abrasiveness, inhomogeneity and anisotropic properties, tool wear rates are inherently high leading to superior cutting forces and detrimental effects on workpiece surface quality and material integrity. Damage such as delamination, cracks or matrix thermal degradation is often observed as the result of uncontrolled tool wear or improper machining conditions. Sensor monitoring of drilling operations is, therefore, highly desirable for process conditions' optimization and tool life maximization. The development of this kind of automated control technologies for process and tool state evaluation can notably contribute to the reduction of scraps and tool costs as well as to the improvement of process productivity in the drilling of CFRP composite material parts. In this paper, multi-sensor process monitoring based on thrust force and torque signal detection and analysis was applied during drilling of CFRP/CFRP laminate stacks for the assembly of aircraft fuselage panels with the scope to evaluate the tool wear state. Different signal-processing methods were utilised to extract diverse types of features from the detected sensor signals. A machine-learning approach based on an artificial neural network (ANN) was implemented to make smart decisions on the timely execution of tool change, which is highly functional for CFRP drilling process automation.

Keywords: stack drilling; CFRP/CFRP laminates; multiple sensor monitoring; tool wear evaluation

1. Introduction

In the aerospace industry, weight reduction is critical to meet environmental requirements (lower emissions) and to reduce management costs (lower fuel consumption). The use of advanced composite materials such as carbon fiber-reinforced plastics (CFRP) is increasing ever more due to their exceptional performance in terms of high specific strength and stiffness, excellent corrosion resistance, and good fatigue resistance [1,2].

In the new generation of aircraft, the percentage of CFRP composite materials has significantly increased, reaching in some cases over 50% of the whole weight of the vehicle. This trend is going to continue in the near future [3].

The assembly of CFRP composite parts is most frequently carried out using mechanical joints, such as rivets, due to the difficulty of realizing welding operations, only applicable to thermoplastic matrix composites, or adhesive joints. Accordingly, drilling is the most widespread CFRP machining process in the aerospace and aeronautical industries. As a matter of fact, in a medium-sized airplane some 85,000 rivets are estimated [4,5]. Drilling of CFRP laminates is frequently carried out manually and cutting tools are often replaced largely before the end of tool life to avoid material damage due to early tool failure [6]. This has a quite negative effect on the cost and productivity of drilling operations.

Nevertheless, drilling of CFRP composite parts is a challenge for manufacturing engineers due to the anisotropic nature of the material, the rapid tool wear caused by abrasive carbon fibers, and the highly concentrated stresses and vibrations. These phenomena may cause critical defects affecting material integrity, surface quality and part acceptability: hole entry and hole exit delamination, geometrical and dimensional errors, interlaminar delamination, fiber pullout, and thermal damage [7–9]. In past decades, diverse non-destructive testing and evaluation techniques were developed and applied to detect defects for quality evaluation of CFRP components [8,10,11].

Delamination, i.e., the separation of laminated layers, is the most critical damage mode since its presence negatively impacts the hole strength and consequently the in-service behaviour of assembled parts [12–14].

The delamination phenomenon is closely influenced by the drilling process parameters [15–20]. In [21,22], an excessively high thrust force is reported as the main process parameter responsible for catastrophic tool failure and extended delamination damage.

Previous studies have shown that delamination usually occurs at hole entry, due to the so called "peel-up" phenomenon, and hole exit, due to the so called "push-out" phenomenon, along the drilled hole periphery [4,23]. The influence of the thrust force for push-out delamination is very high compared to the effect of torque [7].

The optimization of drilling conditions includes the optimal selection of cutting speed and feed rate as well as the thrust force value obtained. Feed rate affects the onset of delamination more severely than spindle speed: in [24], it is shown that minimum peel-up and push-out delamination is verified at minimum feed rate value. However, using too low a feed rate, an increase of contact time between tool and workpiece is generated causing possible thermal damage to the composite material [25].

A higher automation of drilling processes for assembly applications would certainly enhance the productivity of assembly procedures but it needs an effective and dependable drilling process control to ensure hole quality and to fully exploit tool life. A solution can be represented by the development of robust and reliable on-line real-time process control techniques.

In this paper, smart multi-sensor process monitoring based on thrust force and torque signal detection and analysis was applied during drilling of CFRP/CFRP laminate stacks for the assembly of aircraft fuselage panels with the scope to evaluate the tool wear state. Different signal-processing methods were utilised to extract diverse types of features from the detected sensor signals in the time domain, in the frequency domain, and using fractal analysis. A machine-learning approach based on artificial neural network (ANN) data processing was implemented using selected signal features to make smart decisions on the timely execution of tool change, which is highly functional for CFRP drilling process automation, on the basis of tool wear level estimation and tool wear curve reconstruction.

2. Materials and Experimental Procedures

The CFRP/CFRP laminate stacks for the drilling experimental program were made of two overlaid, symmetrical and balanced CFRP laminates of 5 mm thickness (Toray T300 carbon fibers, CYCOM 977-2 epoxy matrix). Each laminate had 26 unidirectional plies with stacking sequence [±452/0/904/0/90/02] s.

A thin 0/90 fabric fiberglass/ epoxy ply was laid on the top and the bottom surfaces of each laminate. The CFRP laminate fabrication was performed by hand layup, vacuum bag moulding, and autoclave cure for 180 min at 180 °C under 6 bar pressure.

In the CFRP/CFRP stack, the laminates were overlaid with their bag sides in contact, representing the severest stack drilling condition (Figure 1). Drilling tests of CFRP/CFRP stacks were performed on a CNC drill press with a clamping system designed to reproduce the industrial drilling operation conditions for assembly of aircraft fuselage panels (Figure 2).

Figure 1. Carbon fiber-reinforced plastic (CFRP)/CFRP laminate stack for drilling experimentation: mold side (smooth surface); bag side (irregular surface).

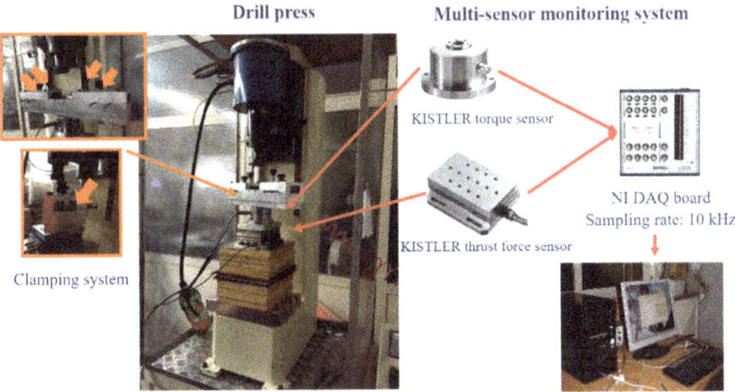

Figure 2. Experimental set-up of CFRP/CFRP stack drilling.

The cutting tools were drill bits made of tungsten carbide with the typical characteristics of twist drills utilised in the aeronautical industry: diameter: D = 6.35 mm; geometry: step/twist; point angle: 120°/125°; helix angle: 20°/30°.

The process conditions utilized for stack drilling were: feed rate = 0.11–0.15–0.20 mm/rev; spindle speed: 2700–6000–9000 rpm (Table 1).

Table 1. Experimental drilling conditions.

	Spindle	Speed	rpm
Feed rate, mm/rev	2700	6000	9000
0.11	X	X	X
0.15	X	X	X
0.20		X	

For each experimental condition, a sequence of 60 holes was realized with the same drill bit.

During the drilling tests, a piezoelectric dynamometer (Kistler 9257) was employed to acquire the thrust force along the z-direction, F_z, and a torque dynamometer (Kistler 9277A25) was utilized to acquire the cutting torque about the z axis, T. As shown in Figure 2, the torque sensor was mounted under the drilling samples clamping system and the thrust force sensor was installed immediately under the torque sensor. The acquired analogue sensor signals were digitalized by a DAQ board (NI USB-6361) at 10 kHz according to the Nyquist–Shannon sampling theorem.

3. Tool Wear Measurement

An optical measuring machine (Tesa Visio V-200, Figure 3) was used to measure the tool flank wear, VB (mm), according to ISO 3685 standard [26]. The drilling tools were fixed on a clamping system (Figure 3) in order to assure the repeatability of the measurement procedure. The VB measurements were carried out at D/6 from the external tool diameter, D, on the left and right cutting edges after every 10 drilled holes (Figure 4).

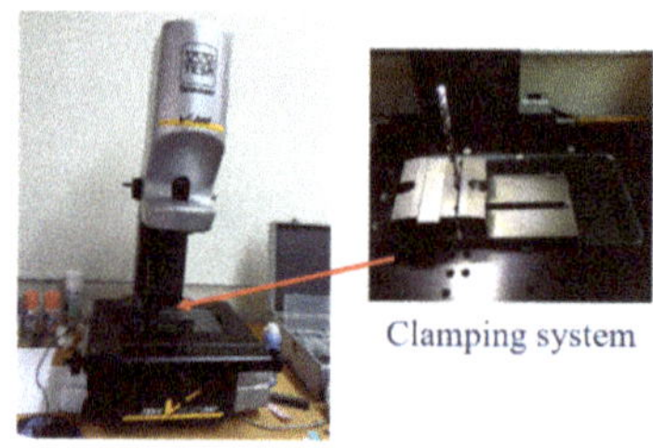

Figure 3. Optical measuring machine and drilling tool clamping system for tool wear measurement.

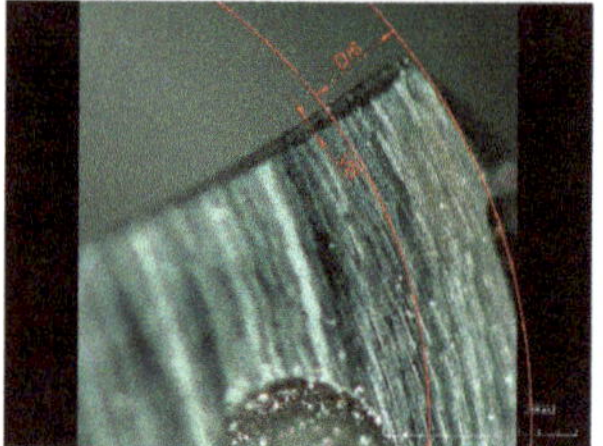

Figure 4. Magnified view of cutting lip and flank wear measurement.

The measured flank wear values were reported in Figure 5 for diverse drilling conditions, highlighting a rising trend in the evolution of tool wear for all operating conditions. A third-order polynomial interpolation of the measured VB values was applied for tool wear curve construction under the different drilling conditions in order to look for correlations between the actual tool wear level and the features extracted from the sensor signals.

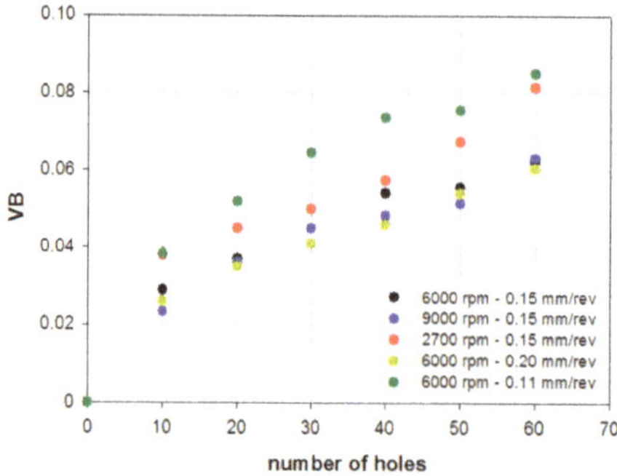

Figure 5. Measured flank wear VB vs. hole number for diverse drilling conditions.

4. Sensor Signal Analysis

4.1. Morphology of Sensor Signals

The thrust force and torque signals acquired during CFRP/CFRP stack drilling are greatly scattered with presence of high-frequency oscillations (Figure 6): this behaviour is strongly related to the anisotropic nature of CFRP laminates [27]. Previous studies on cutting of CFRP materials disclosed that the fiber orientation with respect to the cutting direction determines the mechanism of chip formation and the cut surface quality [28,29]. Thus, different cutting modes can be identified based on the angle formed between cutting edge and reinforcing fibers, also known as fiber cutting angle (Figure 7) [28]. During unidirectional CFRP drilling, the fiber cutting angle varies continuously with drill rotation, determining different mechanical loading and surface quality conditions (Figure 8) [29]. The case of multidirectional CFRP laminate drilling is even more complex: not only the fiber cutting angle varies during drill rotation, but also different cutting modes occur at the same time along the cutting edge, according to the diverse fiber orientations of the multiple plies simultaneously cut by the cutting edge (Figure 9). High-amplitude oscillations in the force and torque signals during drilling of multidirectional CFRP laminates are the sum of multiple waves with a phase difference depending on the different fiber orientations (0°, 45°, 90°, −45°) and an amplitude related to the number of plies with same fiber orientation simultaneously cut by the cutting edge.

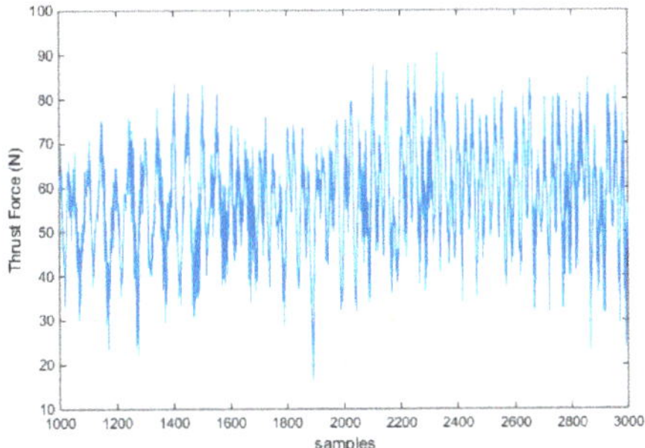

Figure 6. Trust force high frequency signal oscillations (6000 rpm–0.15 mm/rev) [27].

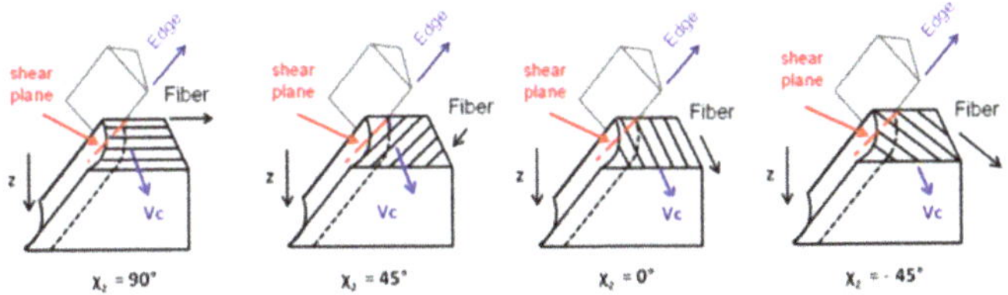

Figure 7. Fiber cutting angle during drilling [28].

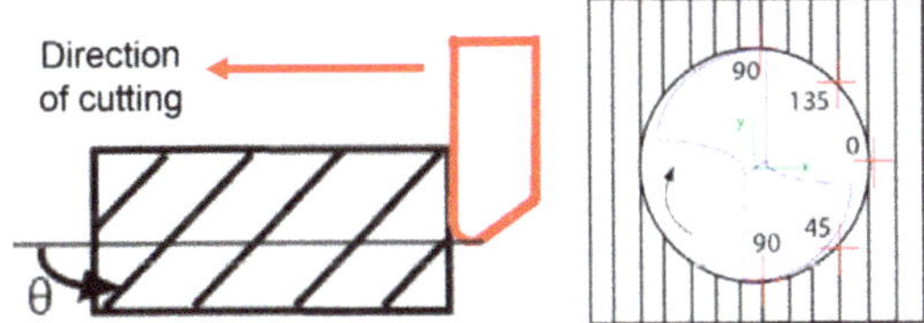

Figure 8. Variation of fiber cutting angle during drilling [29].

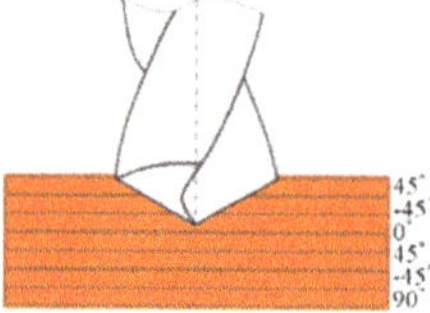

Figure 9. Diverse simultaneous fiber orientations along the cutting edge.

4.2. Sensor Signal Pre-Processing

The raw thrust force and torque sensor signals acquired during the drilling tests included signal portions corresponding to time periods before and after real machining. With the aim to extract sensorial information only when the tool is actually removing material, a signal segmentation procedure was performed on the thrust force signals and synchronically extended to the torque signals. The identification of the start and end of the actual machining portion of the signal was carried out on the basis of thresholds set on the moving average of the thrust force signals.

In Figure 10, the thrust force signal segmentation for hole n° 6 with drilling parameters 6000 rpm and 0.15 mm/rev is reported, showing a drop of thrust force at the interface between the two laminates which are in contact with their bag sides characterized by very irregular surfaces.

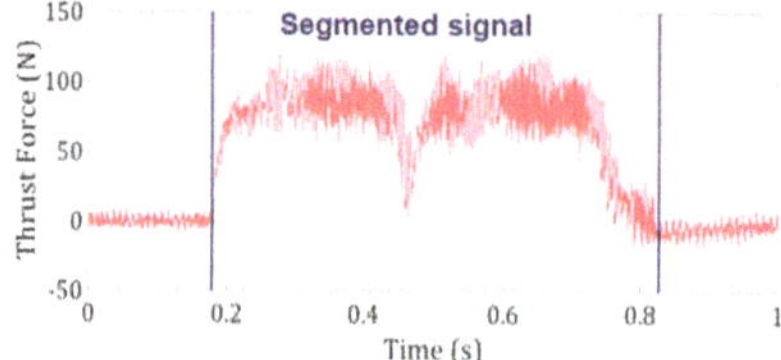

Figure 10. Thrust force segmented signal (6000 rpm–0.15 mm/rev, hole n°6): no real machining occurs before and after the segmented signal [27].

In Figure 11, the thrust force signals are reported versus the increasing number of holes (from 1 to 60) for the operating conditions 2700 rpm–0.15 mm/rev showing that the thrust force and its variance significantly grows with increasing number of holes (ranging from 50 N for hole n.1 to over 200 N for hole n.60). Figure 12 shows the torque signals versus the number of holes for 2700 rpm–0.15 mm/rev drilling conditions highlighting the same previous behaviour.

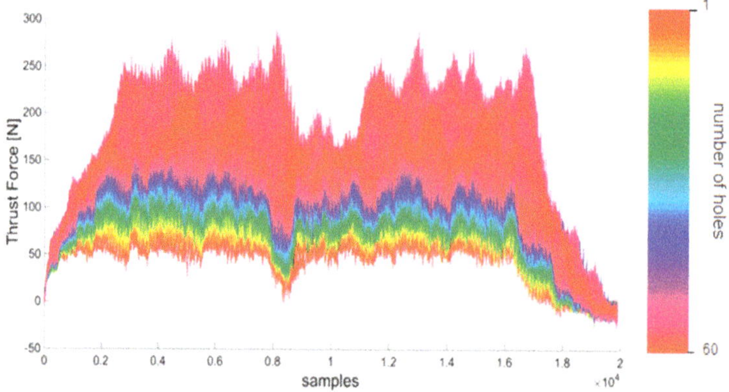

Figure 11. Segmented thrust force signals vs. number of holes (2700 rpm–0.15 mm/rev).

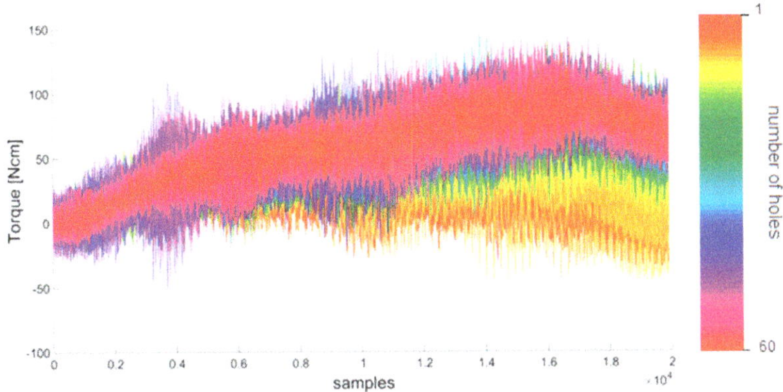

Figure 12. Segmented torque signal vs. number of holes (2700 rpm–0.15 mm/rev).

5. Sensor Signal Feature Extraction

Different signal processing methods were applied to extract diverse types of signal features from the segmented thrust force and torque signals:

- Time domain signal features
- Frequency domain signal features
- Fractal analysis signal features

The extracted signal features were then subjected to selection procedures aimed at identifying those with the highest relation with tool wear level. The selected features were used for the construction of feature pattern vectors (FPV) to be fed as input to ANN-based machine-learning paradigms in order to make decisions on the timely execution of drilling tool change.

5.1. Signals Feature Extraction in the Time Domain

Statistical features in the time domain were extracted from the segmented thrust force and torque signals: arithmetic mean (F_{mean}, T_{mean}), variance (F_{var}, T_{var}), skewness (F_{skew}, T_{skew}), kurtosis (F_{kurt}, T_{kurt}), signal energy (F_{ene}, T_{ene}). Some of these features displayed a good correlation with hole number and thus, expectedly, with tool wear level. In Figure 13, as an example, statistical features extracted from the thrust force signals are plotted for all operating conditions vs hole number up to the last hole n°60. From the figure, it can be noticed that some of the features increase with increasing number of holes while other features decrease with growing hole number. In general, by carrying out a graphical analysis of feature trends, higher or lower degrees of correlation with hole number is verified as a function of the specific statistical feature.

5.2. Signals' Feature Extraction in the Frequency Domain

A fast Fourier transform (FFT) was applied to convert the segmented thrust force and torque signals into the frequency domain. For both thrust force and torque signals, relevant frequency peaks were found corresponding to 1×–2×–3×–4×–5×–6× the drill bit revolution frequency.

Figure 14 show the FFT of thrust force signals for the drilling test performed at 6000 rpm and 0.15 mm/rev, in which the revolution frequency is 6000 rpm/60 = 100 Hz. The highest frequency peaks were observed at 100, 200, 300, 400, 500, 600 Hz, i.e., at 1× 2× 3× 4×–5×–6× the revolution frequency.

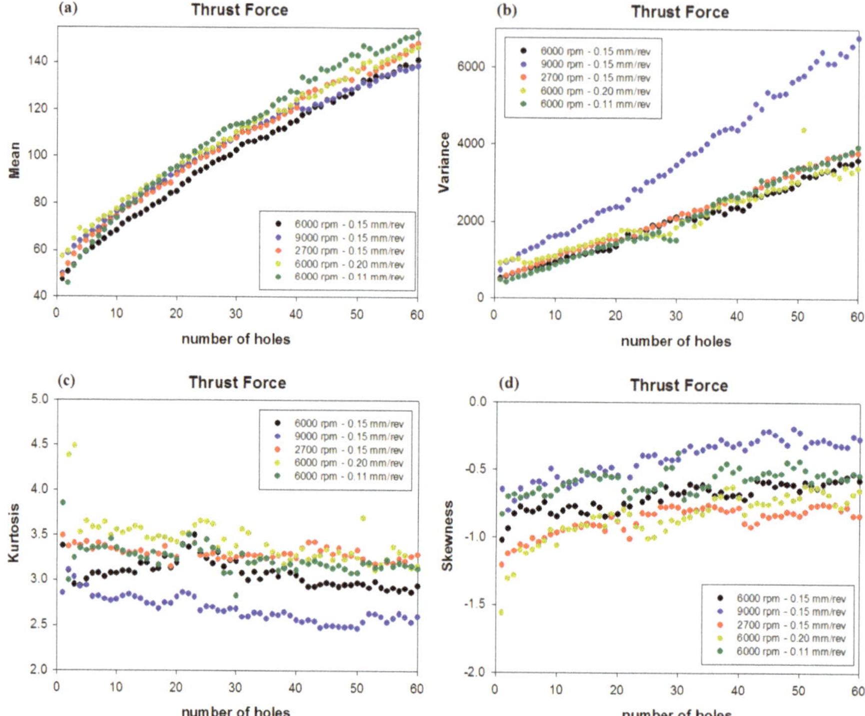

Figure 13. Statistical features extracted from the thrust force signals for all operating conditions vs. hole number: (**a**) arithmetic mean; (**b**) variance; (**c**) kurtosis; (**d**) skewness.

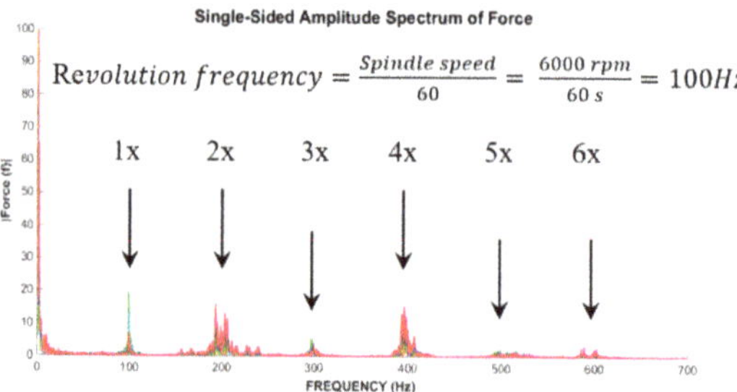

Figure 14. Thrust force signal single-sided amplitude spectrum (6000 rpm–0.15 mm/rev) [27].

In Figure 15, the frequency peaks of the thrust force signal are reported versus hole number for the drilling test performed at 6000 rpm and 0.15 mm/rev. It can be observed that the amplitudes of some of the frequency peaks grow with increasing number of holes, suggesting a potential correlation of the frequency peak amplitude with tool wear progression.

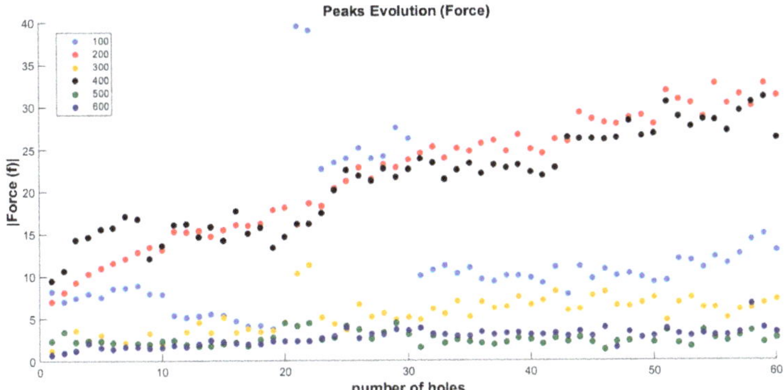

Figure 15. Frequency peaks of the thrust force vs. hole number (6000 rpm–0.15 mm/rev) [27].

5.3. Fractal Analysis Signal Features Extraction

An innovative signal processing approach based on fractal analysis was applied to carry out feature extraction from the segmented thrust force and torque signals.

Fractals were introduced by the mathematician Benoit Mandelbrot to describe the length of Britain's coastline [30]. Nowadays, a fractal is seen as an object which owns a self-affine pattern or singularity. Several applications of fractal analysis were presented in the literature. In metrology, fractal analysis has been investigated and effectively applied to define the roughness of surfaces [31]. As a matter of fact, mathematicians found that rough surfaces have a self-affine behaviour and used the fractal dimension to estimate its roughness [32]. On the other hand, only few research articles show interest in applying this analysis technique in machining, such as for the analysis of sensor signals detected during machining process monitoring.

In this research work, the objective of fractal analysis is to quantify the changes in complexity and shape of the drilling sensor monitoring signals along the tool life. Using regularization fractal analysis, based on convolutions of sensor signals with kernels of different types (rectangular kernel and Gaussian kernel), two sets of curves were built [33]. Two examples of regularization analysis graphs obtained using the rectangular kernel and the Gaussian kernel are displayed in Figure 16 where the different colours refer to the different hole numbers. It can be observed that the regularization curves are influenced by the tool wear level (Figure 16a): they become increasingly bumpy, with higher and more marked arches as the hole number grows.

In order to adapt the analysis to the machining process under study, the regions to extract the fractal parameters were selected based on the tool rotational speed. Two regions, Region 1 and Region 2, were identified in the regularization analysis graphs for the rectangular kernel (Figure 16a) and a third region, Region 3, for the Gaussian kernel (Figure 16b). The boundaries of Regions 1 and 3 were set so as to keep low a values and hold a sufficient number of points for linear regression to extract the slope from the graph curve. The boundaries of region 2 were positioned in order to extract information from the graph arches.

For each region, three fractal parameters were extracted: the slope, D, which represents the fractal dimension quantifying the signal complexity; the y-intercept, G, quantifying the signal ruggedness; and the R-square, R^2, quantifying the auto-scale regularity in the selected region.

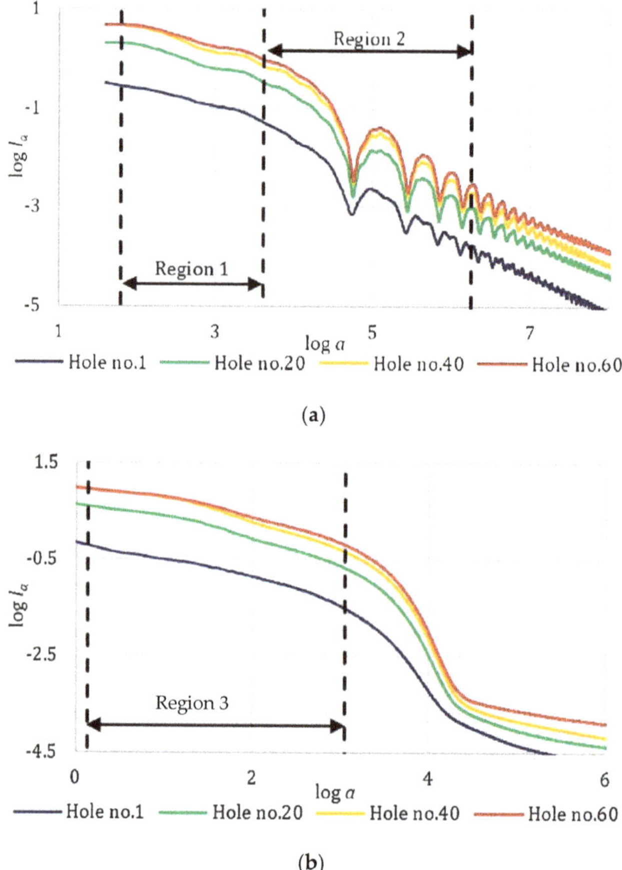

(a)

(b)

Figure 16. Regularization analysis curves: (**a**) rectangular kernel; (**b**) Gaussian kernel [33].

6. Features Selection for Pattern Feature Vector Construction

A statistical approach was employed to evaluate the correlation between segmented time domain, frequency domain and fractal features, on the one hand, and the measured tool flank wear values, on the other hand, using the Spearman correlation coefficient, r_s:

$$r_s = \frac{cov(rg_X, rg_Y)}{\sigma_{rg_X}\sigma_{rg_Y}}$$

where $cov(rg_X, rg_Y)$ is the covariance of the rank variables (i.e., rg_X = features values, rg_Y = tool flank wear values) and $\sigma_{rg_X}\sigma_{rg_Y}$ are the standard deviations of the rank variables.

If $0 < r_s < 0.3$, the correlation is weak; if $0.3 < r_s < 0.7$, the correlation is adequate; if $0.7 < r_s < 1$, a strong correlation occurs. Based on the r_s value, the features displaying the highest correlation with tool wear level were selected as follows:

- Time domain features—Three features were selected from the thrust force signals: thrust force average (F_{mean}), thrust force variance (F_{var}) and thrust force kurtosis (F_{kurt}), and one feature was selected from the torque signal: torque average (T_{mean}). All these featuers displayed a strong correlation with tool wear level.

- Frequency domain features—The following set of features was selected: Fpeak2x, Fpeak4x, Fpeak6x, Tpeak2x, and Tpeak4x. The first three features, extracted from the thrust force signal, displayed the highest correlation with tool wear level.
- Fractal analysis features—Three fractal features were selected from the thrust force signal: the G parameter in region 1 (F_{G1R}), the G parameter in the Gaussian region ($F_{G\text{-}G}$), the Index in region 2 ($F_{Index,2R}$), exhibiting a strong correlation with tool wear. From the torque signal, one fractal feature was selected: the G parameter in the Gaussian region ($T_{G\text{-}G}$), showing an adequate correlation with tool wear level.

The Spearman correlation coefficient was utilized in order to select the features displaying the highest robustness in identifying the tool flank wear level. Accordingly, the selected time domain, frequency domain, and fractal features were used to construct sensor fusion pattern vectors (FPV), containing features from thrust force and torque signals to be then employed for tool wear diagnosis through ANN based machine learning. In particular:

- For time domain, a FPV was constructed containing the selected statistical features of thrust force and torque signals plus the hole number: $FPV_{time} = [F_{mean}, F_{var}, F_{kurt}, T_{mean}, H_n]$
- For frequency domain, a FPV was constructed containing the selected features of thrust force and torque signals: $FPV_{freq} = [Fpeak_{2x}, Fpeak_{4x}, Fpeak_{6x}, Tpeak_{2x}, Tpeak_{4x}]$
- For fractal analysis, a FPV was constructed containing the four selected fractal features of thrust force and torque signals plus the hole number: $FPV_{fract} = [F_{G1R}, F_{G\text{-}G}, F_{Index,2R}, T_{G\text{-}G}, H_n]$

7. Machine Learning Based on Artificial Neural Network (ANN) Data Processing

The constructed time domain, frequency domain and fractal analysis FPV were used as input to ANN based machine learning paradigms for tool wear curve reconstruction aiming at optimal tool life exploitation. Machine learning is an artificial intelligence method based on the idea that machines can learn from data, allowing complex models to be built that can make diagnoses, forecasts or decisions from example data inputs by revealing the patterns embedded in data [34,35]. In this paper, three-layer cascade-forward back propagation ANN were built with diverse configurations, one for each drilling test condition. The architecture of a cascade forward ANN is made of input, hidden and output layers, and comprises connections from the input to each network layer, and from each layer to the successive layers. The input layer collects input patterns while the output layer provides classifications to which input patterns may map [36].

For each drilling test condition, 60 input–output vectors, one for each drilled hole, were formed to set up the ANN learning set. The input layer received in input the selected feature pattern vector (FPV_{time}, FPV_{freq}, or FPV_{fract}) for each drilled hole while the corresponding tool flank wear values, VB, were fed to the output layer during ANN learning.

The utilized ANN had the following architecture:

- the number of input layer nodes was equal to the number of elements of the input FPV, i.e., 5 input nodes;
- the number of hidden layer nodes was set equal to 1x, 2x or 3x the number of input layer nodes, i.e., 5, 10 or 15 hidden nodes;
- the output layer had only one node corresponding to the tool flank wear value, VB.

For ANN training, a Levenberg–Marquardt optimization algorithm [36] was chosen with the following parameters: maximum number of epochs: 1000; performance goal: 0; maximum validation failures: 6; minimum performance gradient: 1×10^{-7}; maximum mu: $1 \times 10^{+10}$; training stop: when the maximum number of epochs is reached and the maximum amount of time is exceeded.

The learning set of 60 FPVs was partitioned into three subsets for training (70%), validation (15%) and testing (15%). The training subset is utilised for calculating the gradient and updating the ANN weights and biases. The validation subset is used to avoid data overfitting by monitoring the error on this subset during training, so that the ANN weights and biases that correspond to the lowest validation error are stored. The testing subset is used to evaluate the performance of the trained ANN. However, the given partition into training, validation and testing subsets can significantly affect the ANN pattern recognition performance in tool wear level evaluation. For this reason, a bootstrap resampling technique was applied to randomly generate the subsets several times with the aim to enhance the estimation of the ANN pattern recognition performance [34,35]. Following the bootstrap procedure, from the original set of 60 FPVs, the training (42 FPV), validation (9 FPV) and testing (9 FPV) subsets were resampled 60 times with replacement. Thus, the overall pattern recognition performance in tool wear level evaluation for tool wear curve reconstruction was estimated by aggregating the recognition rates obtained by all 60 re-samplings.

8. Results and Discussion

The performance of the three-layer cascade-forward backpropagation ANN for smart tool wear curve reconstruction was evaluated using the root mean square error (RMSE):

$$RMSE = \sqrt{\sum_{i=1}^{n} \frac{(\hat{y}_i - y_i)^2}{n}}$$

where $\hat{y}_i$ are the ANN predicted tool flank wear values and y_i are the measured tool flank wear values.

The ANN results obtained reveal that the 5-5-1 ANN configuration provided the lowest RMSE values for smart tool wear curve reconstruction for almost all drilling test conditions in comparison with the 5-10-1 and 5-15-1 ANN configurations. In Table 2, the RMSE values for the 5-5-1 ANN configuration are reported for each drilling condition and for the time domain FPV_{time}, the frequency domain FPV_{freq}, the fractal analysis FPV_{fract}. Moreover, the RMSE values obtained from a fourth FPV, called combined feature pattern vector FPV_{comb}, which includes both time domain and fractal analysis features is a sensor fusion approach, are also reported in the last column of the table.

Table 2. Performance of the 5-5-1 artificial neural network (ANN) configuration expressed as root mean square error (RMSE) in tool wear curve reconstruction for each drilling condition and each constructed feature pattern vector (FPV).

RMSE for ANN Configuration 5-5-1				
Drilling Condition	**ANN Performance in Terms of RMSE for Different FPV**			
(rotational speed-feed rate)	FPV_{time}	FPV_{freq}	FPV_{frac}	FPV_{comb}
2700 rpm–0.11 mm/rev	0.00109	-	0.00045	0.00032
2700 rpm–0.15 mm/rev	0.00407	0.00221	0.00347	0.00321
6000 rpm–0.11 mm/rev	0.00249	0.00195	0.00046	0.00044
6000 rpm–0.15 mm/rev	0.00514	0.00134	0.00076	0.00241
6000 rpm–0.20 mm/rev	0.00061	0.00099	0.00125	0.00035
9000 rpm–0.11 mm/rev	0.00328	-	0.00154	0.00083
9000 rpm–0.15 mm/rev	0.00037	0.00121	0.00051	0.00034
Average RMSE	0.00244	0.00154	0.00120	0.00113

From Table 2 and Figure 17a, it can be noted that the best overall ANN performance was obtained for drilling condition 9000 rpm–0.15 mm/rev with the lowest RMSE equal to 0.00037, whereas the worst overall RMSE, equal to 0.00514, was obtained for drilling condition 6000 rpm–0.15 mm/rev. Both the lowest and the highest RMSE values were obtained when using the time domain feature pattern vector, FPV_{time}.

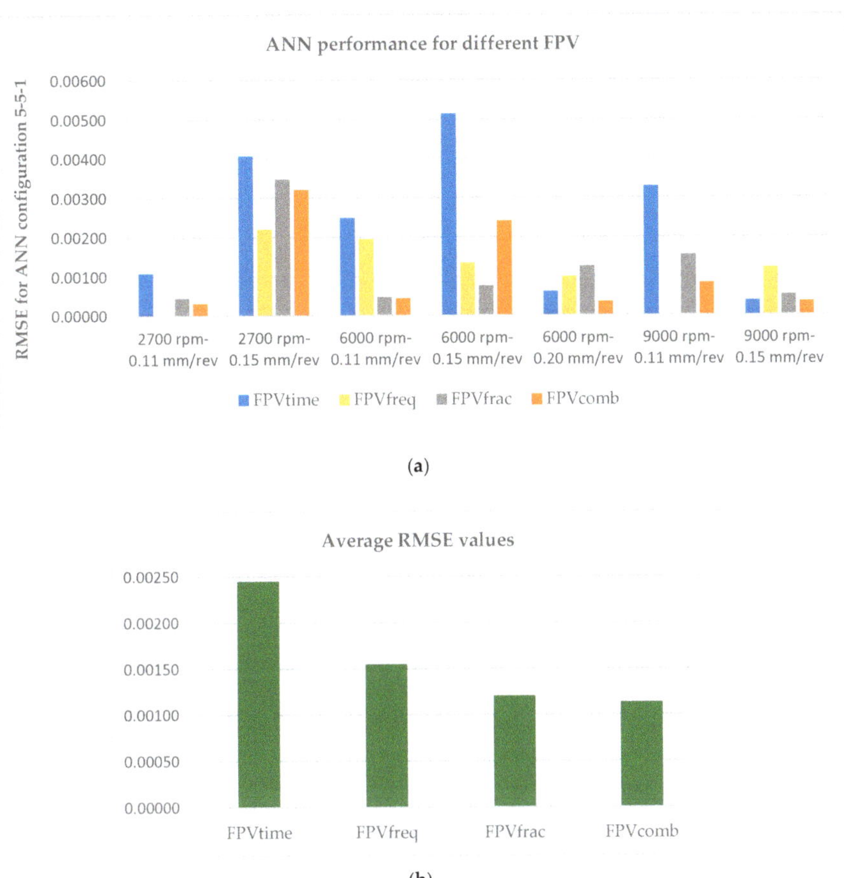

Figure 17. (**a**) RMSE values for the 5-5-1 ANN configuration for each drilling condition and for each constructed FPV; (**b**) RMSE average values for each constructed FPV.

As regards the frequency domain feature pattern vector, FPV_{freq}, the RMSE values ranged between 0.00099 (lowest RMSE) and 0.00221 (highest RMSE).

In the case of the fractal analysis feature pattern vector, FPV_{fract}, the lowest RMSE was equal to 0.00045 whereas the highest RMSE was 0.00347.

The examination of the average RMSE values, evaluated by considering the RMSE values obtained with the same given FPV applied to all the drilling conditions (Table 2 and Figure 17b), indicates that the fractal analysis FPV_{fract} provided the best ANN performance in the classification of tool wear level, while the time domain FPV_{time} came second and the frequency domain FPV_{freq} came third in this ranking.

Thus, to verify the robustness of smart tool wear curve reconstruction by making use of signal features obtained from diverse feature extraction methodologies, the time domain and the fractal analysis features were considered in a combined manner with the scope to construct an additional sensor fusion feature pattern vector, FPV_{comb}. The latter was built using the two most correlated fractal analysis features, the two most correlated statistical features and the hole number: $FPV_{comb} = [F_{G\text{-}G}, F_{Index,2R}, F_{mean}, F_{var}, H_n]$. This 5-component sensor fusion feature pattern vector, FPV_{comb}, was employed for the learning of the same previously utilized 5-5-1 ANN configuration.

In the last column of Table 2, the RMSE values obtained with the 5-5-1 ANN configuration are reported for the case of the combined feature pattern vector FPV$_{comb}$.

The results show that the combination of features extracted via the fractal analysis and time domain methodologies provides the highest robustness in the overall ANN prediction performance for drilling condition 2700 rpm–0.11 mm/rev, yielding a best overall RMSE equal to 0.00032 and an average RMSE equal to 0.00113 (Table 2 and Figure 17b). This confirms the robustness of sensor fusion technology [37] for smart decision making in multi-sensor process monitoring applications [38].

9. Conclusions

Smart multi-sensor monitoring of CFRP/CFRP laminate stack drilling for aeronautical assembly was carried out based on thrust force and torque signal detection and analysis for the on-line evaluation of tool wear state. Different signal-processing methods were utilised to extract and select relevant signal features from sensor signals in the time domain, in the frequency domain, and using fractal analysis. The selected signal features were combined into diverse feature pattern vectors, FPV, and utilised as input in an ANN-based machine-learning paradigm to make smart decisions about the timely execution of tool change on the basis of tool wear level estimation. For all CFRP/CFRP laminate stack drilling conditions, very accurate and robust ANN predictions of residual tool life were achieved utilizing the signal features selected on the basis of their correlation with tool wear level. In particular, the fractal analysis and the time domain FPVs provided the best ANN performance in the classification of tool wear level. Therefore, to verify the robustness of smart tool wear curve reconstruction, sensor fusion pattern vectors obtained combining features extracted via fractal analysis and time-domain methodologies were constructed and employed for ANN learning and testing. The best overall ANN prediction was achieved by these FPVs confirming the robustness of sensor fusion technology for smart decision making in multi-sensor process-monitoring applications.

The utilized smart multi-sensor monitoring procedure can thus be reliably employed for on-line tool wear diagnosis aimed at end-of-tool-life dependable forecast required to implement an effective process automation in composite material parts assembly based on mechanical drilling. In this framework, a condition-based tool replacement strategy can be adopted instead of a time-based strategy, accomplishing the enhancement of productivity together with the reduction of scrap rate and tool cost in CFRP/CFRP stack-drilling processes for aeronautical assembly.

Author Contributions: Conceptualization, R.T.; methodology, A.C.; validation, T.S.; investigation, L.N.; data curation, T.S.; supervision, R.T. All authors have read and agreed to the published version of the manuscript.

Funding: This research received no external funding.

Acknowledgments: The Fraunhofer Joint Laboratory of Excellence on Advanced Production Technology (Fh J_LEAPT UniNaples) at the Department of Chemical, Materials and Industrial Production Engineering, University of Naples Federico II, is gratefully acknowledged for its contribution and support to this research activity.

Conflicts of Interest: The authors declare no conflict of interest.

References

1. M'Saoubi, R.; Axinte, D.; Soo, S.L.; Nobel, C.; Attia, H.; Kappmeyer, G.; Engin, S.; Sim, W.-M. High performance cutting of advanced aerospace alloys and composite materials. *CIRP Ann. Manuf. Technol.* **2015**, *64*, 557–580. [CrossRef]
2. Teti, R. Machining of composite materials. *CIRP Ann. Manuf. Technol.* **2002**, *51*, 611–634. [CrossRef]
3. Rajak, D.K.; Pagar, D.D.; Menezes, P.L.; Linul, E. Fiber-reinforced polymer composites: Manufacturing, properties, and applications. *Polymers* **2019**, *11*, 1667. [CrossRef] [PubMed]
4. Liu, D.; Tang, Y.; Cong, W.L. A review of mechanical drilling for composite laminates. *Compos. Struct.* **2012**, *94*, 1265–1279. [CrossRef]
5. Davim, J.P. *Machining Composites Materials*; John Wiley & Sons: Hoboken, NJ, USA, 2013.
6. Brinksmeier, E.; Fangmann, S.; Rentsch, R. Drilling of composites and resulting surface integrity. *CIRP Ann. Manuf. Technol.* **2011**, *60*, 57–60. [CrossRef]

7. Jinyang, X.; Chao, L.; Sipei, M.; Qinglong, A.; Ming, C. Study of drilling-induced defects for CFRP composites using new criteria. *Compos. Struct.* **2018**, *201*, 1076–1087.

8. Geng, D.; Liu, Y.; Shao, Z.; Lu, Z.; Cai, J.; Li, X.; Jiang, X.; Zhang, D. Delamination formation, evaluation and suppression during drilling of composite laminates: A review. *Compos. Struct.* **2019**, *216*, 168–186. [CrossRef]

9. Caprino, G.; Tagilaferri, V. Damage developments in drilling glass fiber reinforced plastics. *Int. J. Mach. Tools Manuf.* **1995**, *35*, 817–829. [CrossRef]

10. Cheng, L.; Tian, G.Y. Surface crack detection for carbon fiber reinforced plastic (CFRP) materials using pulsed eddy current thermography. *IEEE Sens. J.* **2011**, *11*, 3261–3268. [CrossRef]

11. Yi, Q.; Tian, G.Y.; Malekmohammadi, H.; Zhu, J.; Laureti, S.; Ricci, M. New features for delamination depth evaluation in carbon fiber reinforced plastic materials using eddy current pulse-compression thermography. *NDT E Int.* **2019**, *102*, 264–273.

12. Fleischer, J.; Teti, R.; Lanza, G.; Mativenga, P.; Möhring, H.; Caggiano, A. Composite materials parts manufacturing. *CIRP Ann. Manuf. Technol.* **2018**, *67*, 603–626. [CrossRef]

13. Al-Wandi, S.; Ding, S.; Mo, J. An approach to evaluate delamination factor when drilling carbon fiber-reinforced plastics using different drill geometries: Experiment and finite element study. *Int. J. Adv. Manuf. Technol.* **2017**, *93*, 4043–4061. [CrossRef]

14. Geier, N.; Davim, J.P.; Szalay, T. Advanced cutting tools and technologies for drilling carbon fibre reinforced polymer (CFRP) composites: A review. *Compos. Part A Appl. Sci. Manuf.* **2019**, *125*, 105552. [CrossRef]

15. Melentiev, R.; Priarone, P.C.; Robiglio, M.; Settineri, L. Effects of tool geometry and process parameters on delamination in CFRP drilling: An overview. *Procedia CIRP* **2016**, *45*, 31–34. [CrossRef]

16. Franke, V. Drilling of long fiber reinforced thermoplastics—Influence of the cutting edge on the machining results. *CIRP Ann.* **2011**, *60*, 65–68. [CrossRef]

17. Shyha, I.; Soo, S.L.; Aspinwall, D.; Bradley, S. Effect of laminate configuration and feed rate on cutting performance when drilling holes in carbon fibre reinforced plastic composites. *J. Mater. Process. Technol.* **2010**, *210*, 1023–1034. [CrossRef]

18. Geier, N.; Szalay, T.; Takács, M. Analysis of thrust force and characteristics of uncut fibres at non-conventional oriented drilling of unidirectional carbon fibre-reinforced plastic (UD-CFRP) composite laminates. *Int. J. Adv. Manuf. Technol.* **2019**, *100*, 3139–3154. [CrossRef]

19. Lissek, F.; Tegas, J.; Kaufeld, M. Damage quantification for the machining of CFRP: An introduction about characteristic values considering shape and orientation of drilling-induced delamination. *Procedia Eng.* **2016**, *149*, 2–16. [CrossRef]

20. Caggiano, A.; Napolitano, F.; Nele, L.; Teti, R. Multiple sensor monitoring for tool wear forecast in drilling of CFRP/CFRP stacks with traditional and innovative drill bits. *Procedia CIRP* **2018**, *67*, 404–409. [CrossRef]

21. Hocheng, H.; Tsao, C.C. Comprehensive analysis of delamination in drilling of composite materials with various drill bits. *J. Mater. Process. Technol.* **2003**, *140*, 335–339. [CrossRef]

22. Saoudi, J.; Zitoune, R.; Mezlini, S.; Gururaja, S.; Seitier, P. Critical thrust force predictions during drilling: Analytical modeling and X-ray tomography quantification. *Compos. Struct.* **2016**, *153*, 886–894. [CrossRef]

23. Hocheng, H.; Tsaom, C.C. The path towards delamination-free drilling of composite materials. *J. Mater. Process. Technol.* **2005**, *167*, 251–264. [CrossRef]

24. Ahmad, J. *Machining of Polymer Composites*; Springer: Boston, MA, USA, 2009.

25. Rawat, S.; Attia, H. Characterization of the dry high speed drilling process of woven composites using Machinability Maps approach. *CIRP Ann. Manuf. Technol.* **2009**, *58*, 105–108. [CrossRef]

26. *Tool-Life Testing with Single-Point Turning Tools*; ISO 3685:1993; International Organization for Standardization (ISO): Geneva, Switzerland, 1993.

27. Caggiano, A.; Centobelli, P.; Nele, L.; Teti, R. Multiple sensor monitoring in drilling of CFRP/CFRP stacks for cognitive tool wear prediction and product quality assessment. *Procedia CIRP* **2017**, *62*, 3–8. [CrossRef]

28. Bonnet, C.; Poulachon, G.; Rech, J.; Girard, Y.; Costes, J.P. CFRP drilling: Fundamental study of local feed force and consequences on hole exit damage. *Int. J. Mach. Tools Manuf.* **2015**, *94*, 57–64. [CrossRef]

29. Karpat, Y.; Bahtiyar, O.; Deger, B.; Kaftanoğlu, B. A mechanistic approach to investigate drilling of UD-CFRP laminates with PCD drills. *CIRP Ann. Manuf. Technol.* **2014**, *63*, 81–84. [CrossRef]

30. Mandelbrot, B.B. *The Fractal Geometry of Nature*; W. H. Freeman: San Francisco, CA, USA, 1982; p. 57.

31. Russ, J.C. Fractal geometry in engineering metrology. In *Metrology and Properties of Engineering Surfaces*; Mainsah, E., Greenwood, J.A., Chetwynd, D.G., Eds.; Springer: Boston, MA, USA, 2001.

32. Saghi, Z.; Xu, X.; Möbus, G. Three-dimensional metrology and fractal analysis of dendritic nanostructures. *Phys. Rev.* **2008**, *78*, 205428. [CrossRef]
33. Caggiano, A.; Rimpault, X.; Teti, R.; Balazinski, M.; Chatelain, J.F.; Nele, L. Machine learning approach based on fractal analysis for optimal tool life exploitation in CFRP composite drilling for aeronautical assembly. *CIRP Ann.* **2018**, *67*, 483–486. [CrossRef]
34. Alpaydin, E. *Introduction to Machine Learning*; MIT Press: Cambridge, MA, USA, 2014.
35. Bishop, C.M. *Pattern Recognition and Machine Learning*; Springer: New York, NY, USA, 2006.
36. Jordan, M.I.; Bishop, C.M. Neural Networks. In *Computer Science Handbook*, 2nd ed.; Tucker, A.B., Ed.; Chapman & Hall/CRC Press LLC: Boca Raton, FL, USA, 2004.
37. Mitchell, H.B. *Multi-Sensor Data Fusion*; Springer: Berlin/Heidelberg, Germany, 2007.
38. Segreto, T.; Teti, R. Machine learning for in-process end-point detection in robot-assisted polishing using multiple sensor monitoring. *Int. J. Adv. Manuf. Technol.* **2019**, *103*, 4173–4187. [CrossRef]

Article

Cutting Speed and Feed Influence on Surface Microhardness of Dry-Turned UNS A97075-T6 Alloy

Sergio Martín-Béjar, Francisco Javier Trujillo Vilches *, Carolina Bermudo Gamboa and Lorenzo Sevilla Hurtado

Department of Civil, Material and Manufacturing Engineering, EII, University of Malaga, C/Dr, Ortiz Ramos s/n, E-29071 Malaga, Spain; smartinb@uma.es (S.M.-B.); bgamboa@uma.es (C.B.G.); lsevilla@uma.es (L.S.H.)
* Correspondence: trujillov@uma.es; Tel.: +34-951-952-245

Received: 23 December 2019; Accepted: 23 January 2020; Published: 5 February 2020

Abstract: In this work, an analysis of the cutting speed and feed influence on surface roughness and microhardness of UNS A97075-T6 alloy, turned under dry conditions, was carried out. The results were compared before and after a corrosion process. The influence of these cutting parameters on each of these variables was analyzed, as well as the possible interrelation between them. The microgeometrical deviations showed a general trend to increase with feed. However, no significant modifications were observed as a function of the cutting speed. This trend was softer after the corrosion process, due to the surface alterations produced by pitting corrosion, which resulted in higher dispersion of the experimental data. In addition, a surface microhardness increment was observed in all samples, after machining and before corrosion, regardless of the cutting parameter values. The experimental results revealed that the mechanical effects, produced by the feed, should not be neglected against the thermal effects, produced by the cutting speed, within the range of the tested cutting speed. Finally, the corrosion process negatively affected the microhardness, but it was not possible to establish a direct relationship between the cutting parameters, surface roughness, and microhardness after a corrosion process.

Keywords: dry turning; surface roughness; microhardness; UNS A97075; aluminum alloys

1. Introduction

Pure aluminum or aluminum with a very low weight percentage of alloying elements (1000 series) has, among its mechanical properties, a low wear resistance and a low hardness. Therefore, the addition of alloying elements implies an improvement of the aluminum surface characteristics and mechanical properties, making these alloys one of the most used materials in the industrial activity [1]. One of the industrial sectors in which aluminum alloys are of great interest is the aeronautical industry. Their excellent density and mechanical properties make this type of material suitable for the manufacturing of aircraft structural elements. In particular, the 2000 (Al–Cu) and 7000 (Al–Zn) series are widely used for structural elements in the wings, the fuselage, or the tail of the aircraft [2]. These structural components are usually high-rigidity elements and, frequently, they need to be manufactured in one piece. In these cases, the machining operations are usually quite competitive from a productive point of view. In particular, milling operations are used for large pocketing operations in rib-shaped elements, drilling operations are frequent in the machining of holes for rivets, and turning operations are used to manufacture coupled parts with cylindrical shapes [3].

Machining operations result in alterations in the surface of the stock material. These alterations not only affect their geometrical characteristics (at macro and micro scale) but also modify the physical–chemical and mechanical properties of the surface (residual stresses, microstructural alterations, cracks and microcracks, microhardness, corrosion resistance, etc.) [4–6]. All these alterations (both material properties and geometry) are part of the surface integrity concept [7–9], which is one of the most valued aspects from the point of view of quality requirements, especially in structural elements for aircraft, where these requirements are very high due to functionality and reliability reasons [10].

Lubricants or coolants, applied during the machining process (cutting fluids), are traditionally used to increase the productivity of these processes and the machined part functionality. However, the current trend regarding the use of sustainable manufacturing technologies led to a reduction in the use of these highly polluting substances. This resulted in the application of more efficient techniques, such as MQL (minimum quantity lubricant), or the total suppression of these substances (dry machining) [11–15]. In addition, aluminum alloys are frequently used hybridized with other materials, such as carbon fiber-reinforced polymers (CFRP) or titanium, to form structural elements, such as fiber metal laminates (FML) [16]. In those cases, the chip recycling (mixture of metal, carbon fiber, and cutting fluids) becomes complex and expensive. In addition, it should be pointed that CFRPs do not exhibit good behavior under the cutting fluid action [17].

However, in dry machining, the thermal effects become more relevant compared to the effects of plastic deformation due to mechanical actions on the surface. This results in greater surface alterations at both microstructural and geometric levels [18–20]. Due to the high importance of the surface state in phenomena such as crack and microcrack initiation, which can affect significant properties such as fatigue resistance or corrosion [21], it is necessary to analyze the effect of the machining input variables on the machined part surface integrity.

Thus, the surface microhardness of aircraft structural components is particularly relevant, given their influence on the bearing capacity and the wear resistance of these elements [22,23]. In this sense, the cutting parameters are some of the most important machining input variables to keep in mind regarding this output variable. Although there are numerous works in the literature that analyzed the influence of cutting parameters on surface microhardness [24–26], these studies should be completed in the case of dry machining of aluminum alloys.

Thus, Kurkute et al. [27] analyzed the feed and cutting speed influence on the cutting forces and the surface roughness of the UNS A96063 alloy, in dry-turning operations, in order to obtain the optimal machining conditions for low surface roughness (*Ra*) and high surface microhardness (*HV*). Cutting speed values between 10 and 50 m/min and feed values between 0.40 and 0.80 mm/r were used. A lower roughness and higher microhardness were obtained for intermediate feed (0.5 mm/r) and high cutting speed values (40 m/min). In addition, a potential model was developed in this work. That model allowed obtaining the microhardness as a function of the feed rate, the cutting speed, the cutting forces, and the number of machining steps (roughing and finishing).

Surya et al. [28] analyzed the cutting speed and cutting depth influence on the microhardness of dry-turned UNS A97075 alloy. Different cutting speed (50, 100, 200 and 300 m/min) and cutting depth values (0.3, 0.6 and 0.9 mm) were used, whereas the feed remained constant (0.05 mm/r). In this case, a surface microhardness decrease was observed when the cutting speed was increased. In addition, a reduction in the microhardness was also observed when the cutting depth was increased, but its influence was less noticeable. The authors attributed this behavior to the heat generation and dissipation on the machined surface.

As previously mentioned, the machining operation modifies the surface microstructure of the machined parts. In this regard, Rotella et al. [29] used microhardness measurements to evaluate the depth reached by these microstructural alterations, in the dry turning of UNS A97075-T651 alloy. Several hardness tests were carried out for different surface layer depths (up to 500 μm). A microhardness reduction was observed as a function of the depth, and it was finally stabilized at greater depths. The cutting speed and the tool-tip radius influence was also analyzed. The results obtained showed that higher cutting speed values resulted in greater microstructure modifications, with a reduction in the grain size. In addition, cutting tools with higher tool-tip radius generated a greater reduction in the grain size and higher values of surface microhardness, as a consequence of the contact surface reduction between the tool and the machined part.

Similarly, Campbell et al. studied the microstructure modifications as a function of the surface microhardness, in the dry turning of UNS A97075-T651 alloy [30]. In this case, the turning operations were carried out with a constant feed (0.076 mm/r) and cutting speed values between 360 and 720 m/min, while also varying the tool rake angle (0°–15°). The authors indicated that the cutting speed increments and low tool rake angle generated a microstructure modification at higher surface depths, as a consequence of the temperature increment in the cutting area. Contrary to the results obtained by Rotella et al. in [29], Campbell [30] stated that the microhardness in a closer area to the surface decreased until 50-μm depth. Once this value was reached, the microhardness tended to stabilize at higher values than the initial ones and, therefore, a more homogeneous microstructure was obtained.

With regard to the UNS A96061 alloy, Akkurt et al. [26] performed different conventional and non-conventional machining tests, in order to study the microhardness alterations induced by the process in the part surface. It was observed that, in all tested machining processes, the microhardness decreased in the most superficial layer of the machined part, compared to the starting material. This microhardness reduction is justified by the thermal effect of the machining operation on the machined surface. Moreover, this thermal effect is more important in surfaces that present greater irregularities and, therefore, less superficial microhardness.

On the other hand, 7000 aluminum alloy series have lower corrosion resistance compared to pure aluminum. Its main alloying element is Zn (~6%), which makes the appearance of the alumina protective layer (with higher corrosion resistance) more difficult [31]. For the 2000 and 7000 series, corrosion is usually localized, with the appearance of pitting common, as well as intergranular corrosion, which can be considered as the generation and nucleation points of microcracks that affect the continuity of the surface microstructure [32–34].

Despite the lack of research studying the cutting parameter influence on corrosion behavior of machined parts, several studies revealed that low cutting speed and high feed values result in machined parts with lower corrosion resistance [35–37]. In addition, the surface microhardness has a close connection with corrosion resistance. The metallic material corrosion resistance depends essentially on the ability to form passive bonded layers and its mechanical properties. Usually, higher hardness and tensile strength can be related to higher corrosion resistance [36,38,39].

Yue et al. [40] studied the effect of the immersion corrosion process (saline solution 3.5% NaCl) in parts of the UNS A92009 alloy, machined using three different techniques: turned with ceramic tool, turned with diamond tip tool, and WEDM (wire electrical discharge machining). The results revealed that the samples machined by WEDM showed less corrosion resistance than the turned ones. The authors justified this result due to the greater surface irregularity obtained after WEDM machining. In addition, turned parts with the diamond tool were less affected by the corrosion process.

Welcome at al. [41] evaluated the cutting parameters influence on the corrosion behavior of the UNS A92024-T3 aluminum alloy. After machining, the specimens were submerged in a saline solution (3.5% NaCl). Then, the surface finish after corrosion was compared with the initial one (after machining). The results showed that the use of lower feed and higher cutting speed resulted in a better corrosion resistance, as a consequence of the better surface quality of the machined parts.

As shown in the literature review, there are several studies that related the cutting parameters to surface properties such as microhardness and corrosion resistance in dry-machined aeronautical aluminum alloys. However, few studies included the feed as a parameter of influence. This may be logical, keeping in mind that most of these studies were carried out at high cutting speed values (above 100 m/min), where thermal effects (mainly influenced by the cutting speed) predominate over mechanical effects, in which the feed has a greater influence. However, as previously mentioned, these alloys are frequently used hybridized with other materials, such as CFRP and Ti, which require the application of lower cutting speeds. In addition, in no case were the synergistic effects between cutting parameters, geometric alterations, microhardness, and corrosion resistance analyzed.

Therefore, an analysis of cutting speed and feed influence on surface roughness and microhardness of UNS A97075-T6 alloy, turned under dry conditions, was carried out in this work. The results were compared before and after a corrosion process. The influence of these cutting parameters on each of these variables (surface roughness, microhardness, and corrosion resistance) was analyzed, as well as the possible interrelation between them.

2. Materials and Methods

Several dry-turning tests were carried out in order to evaluate the cutting parameter influence on the surface microhardness of the aluminum alloy UNS A97075-T6. The tested alloy composition (% mass), obtained by arc atomic emission spectroscopy (AES), is shown in Table 1.

Table 1. Evaluated aluminum–zinc alloy UNS A97075-T6 composition (% weight).

Zn	Mg	Cu	Cr	Si	Mn	Al
6.03	2.62	1.87	0.19	0.09	0.07	Balance

Cylindrical bars (diameter, D = 20 mm; length, L = 110 mm) were used as stock geometry to manufacture the specimens. The sample final geometry is shown in Figure 1a. A step (25-mm length) was machined at the end of the specimen, in order to ensure a proper grip to the chuck (Figure 1b). A length of 65 mm was turned as the microhardness measurement area. The starting diameter was reduced to 17 mm in two stages. Firstly, a roughing operation was performed to reach 18 mm; then, a finishing operation was carried out to reduce the diameter to 17 mm. A new tool was used in this last operation, to ensure the same initial conditions. The machining operations were carried out in a CNC turning center.

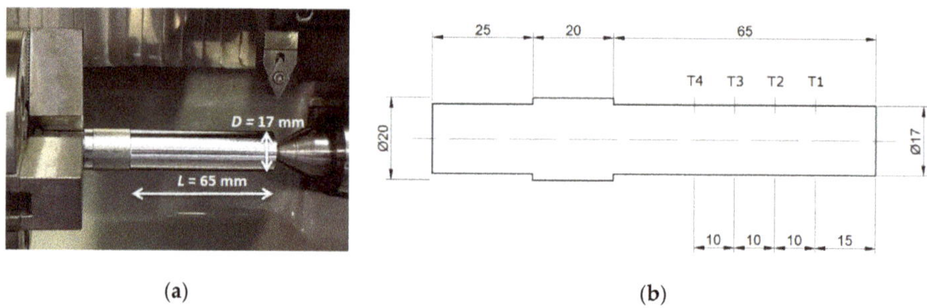

Figure 1. (a) Sample geometry; (b) measurement area (all dimensions in mm).

Different cutting parameter values were used to perform the finishing step. Their values are shown in Table 2. The cutting depth (a_p) remained constant in every test. To evaluate the thermo-mechanical effects of the cutting parameters on the surface micro-hardness, several values of cutting speed (v_c) and feed (f) were used. As previously commented, it is necessary to emphasize that, although this alloy is not usually machined with low cutting speed values, its use hybridized with other materials forces the use of low values. The application of this cutting speed range, together with the analysis of the feed influence, is one of the main novelties of this work.

Table 2. Cutting conditions.

v_c (m/min)	f (mm/r)	a_p (mm)
40	0.05	
60	0.10	1
80	0.15	
	0.20	

Hence, 12 samples were machined (one for each cutting parameter combination). The used cutting tools were uncoated WC-Co inserts, with ISO reference DCMT 11T308-14 IC, a tool-tip angle of 55°, and a tip radius of 0.8 mm.

In order to analyze the surface roughness influence on microhardness (before and after a corrosion process), the roughness profile was obtained. It is necessary to point out that the evolution of the microgeometrical deviations as a function of the cutting parameters, in the dry machining of this alloy, was widely studied. Nevertheless, in this work, the microgeometrical deviations were measured with a double purpose. On one hand, they may be used as an element of control and comparison with previous research. On the other hand, they could be useful to analyze possible synergistic actions between cutting parameters, hardness, surface roughness, and corrosion behavior. For this purpose, a Mitutoyo portable roughness tester, model Mitutoyo SURFTEST SJ-210, was used. The average roughness (Ra) and the maximum height of the profile (Rz) were used as parameters to evaluate the surface quality (Equations (1) and (2)).

$$Ra = \frac{1}{lr} \int_0^{lr} |Z(x)|dx, \tag{1}$$

$$Rz = Zp + Zv, \tag{2}$$

where lr is the sampling length, $Z(x)$ is the profile height, Zp is the peak height, and Zv is the peak valley of the roughness profile.

Two measures were obtained along two different areas. The measurement was repeated along four specimen generatrix (G1–G4, 90° apart; Figure 2). Therefore, eight roughness profiles were obtained for each specimen. The surface roughness measurement was carried out according to the UNE-EN ISO 4288: 1998 standard [42]. The used parameters are indicated below.

- Filter: normal Gaussian distribution;
- Measurement or scan length (ls): 2.5 μm;
- Basic or contact length (lc): 0.8 mm.

Figure 2. Roughness profile measurement set-up.

A Vickers hardness test was performed on each specimen, according to the UNE-EN ISO 6507-1: 2018 standard [43]. A non-machined specimen (stock bar) was used as a control element, to compare the results with the machined samples. The applied load was 0.05 kgf (0.05 HV), corresponding to a value of 0.49 N. The tests were carried out on two different generatrices, obtaining hardness results at four points (T1, T2, T3, and T4) on the same generatrix (Figure 1b). The hardness tests were carried out by using a MATSUZAWA MXT70 micro-hardness tester, with a pyramid-shaped indenter (with a square diamond base and an angle between opposite faces of 136°; Figure 3). The temperature of the measurement area was controlled, remaining within the standard limits of 23 ± 5 °C [43].

Once the surface hardness was measured, the corrosion process for the machined and non-machined samples was carried out. The samples were placed by immersion within a corrosive environment. For this purpose, a solution of deionized water and NaCl (3.5% concentration) was used, keeping the specimens in that environment for 72 h.

During the corrosion process, the corrosive medium temperature was continuously checked, remaining at 26 ± 1 °C. In addition, a pump was placed in the solution to keep the saline medium moving and to avoid NaCl stratification, ensuring the corrosion process homogeneity on the surface of all samples. After the corrosion process, the surface roughness profile was measured in every sample (following the same methodology). After that, the hardness tests were carried out once again, under the same conditions previously described.

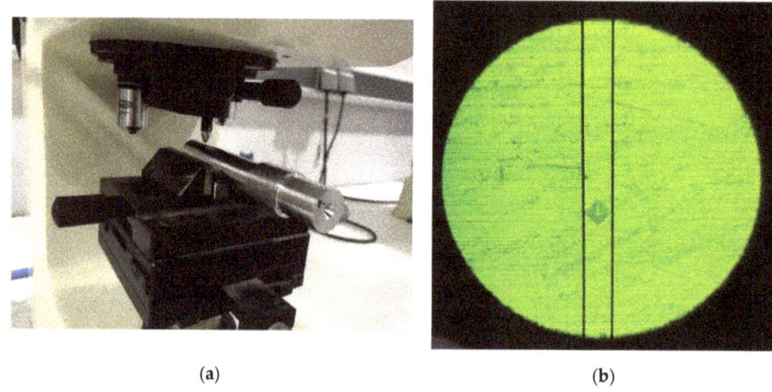

(a)   (b)

Figure 3. (**a**) Hardness test set-up; (**b**) generated mark in the test point.

Finally, scanning electron microscopy (SEM) and energy-dispersive spectroscopy (EDS) techniques were used in order to analyze the alloy microstructure and micro-composition, before and after machining. An etchant (2.5% HNO_3, 1.5% HCl, 1% HF; 95% H_2O) was used to show the grain structure. An image processing software was used to measure the grain size, according to the standard ASTM E112 "Standard Test Methods for Determining Average Grain Size" (Equations (3)–(5)).

$$N = 2^n A^{-1}, \tag{3}$$

$$n = n_A + Q, \tag{4}$$

$$Q = 2log_2\left(\frac{\mu}{100}\right), \tag{5}$$

where N is the number of grains, n is the grain size, n_A is the apparent grain size, Q is the correction factor, and μ is the micrograph magnification.

3. Results and Discussion

Figures 4 and 5 plot the evolution of Ra and Rz, respectively, as a function of the cutting parameters, v_c and f, before (Figures 3a and 4a) and after (Figures 3b and 4b) the corrosion process. These values were calculated as the average values of the experimental results, measured in four different lines along the specimens. Additionally, the Ra and Rz values of the non-machined sample were identified in the figures, in order to analyze the induced alterations by the machining process.

The results revealed that the feed was the most influential variable on surface roughness. Both parameters, Ra and Rz, showed a general trend to increase with f, regardless of v_c. This trend was more noticeable from $f = 0.10$ to 0.20 mm/r. For samples before corrosion, Ra was three times higher for

$f = 0.10$ mm/r than for 0.20 mm/r. For samples after corrosion, the effect of f was amplified, reaching four-fold higher values for 0.20 mm/r than for 0.1 mm/r. The feed influence was less evident in the range of low f applied (0.05–0.10 mm/r). For these low f values, *Ra* and *Rz* showed similar values to samples before machining.

With regard to v_c, no significant modifications were observed in *Ra* or *Rz* when this parameter was modified. Therefore, the effect of the natural roughness, induced by the feed, was predominant over the thermal effects that cutting speed variations may have caused during the cutting process, such as the appearance of built-up edge (BUE), which may have affected the surface quality [5,44]. In the dry machining of this alloy, this indirect adhesion tool wear is usually a consequence of the thermo-mechanical effects at high feed. Nevertheless, the effect of cutting speed becomes more relevant for low feed, giving rise to BUE with higher intensity for higher cutting speed values. However, the low cutting speed applied in this research made this effect less noticeable and, as a result, the cutting speed effect on surface roughness was almost negligible [45–47].

It is necessary to point out that all these observations are in good agreement with previous research, in connection with the parametric analysis of surface roughness in the dry turning of the 2000 and 7000 aluminum alloys series [45,47,48].

Regarding the corrosion process, no significate changes in *Ra* were observed. However, the dispersion in the results seemed to be a little bit higher. This fact was more evident for *Rz*. These results may be explained considering that *Ra* was calculated as an average value of peaks and valleys of the roughness profile, while *Rz* was obtained as the maximum height of the profile. The appearance of pitting and intergranular corrosion, mainly in the inter-metallic elements ($Al_{23}CuFe_4$), resulted in surface irregularities that may have affected the values of the microgeometrical deviations at one point (Figure 6). This effect was softened in the calculation of average values to obtain *Ra*.

In addition, an increment in *Rz* was observed after the corrosion process, regardless of v_c and f. Only for the highest f (0.20 mm/r) was this increment slightly higher. A poorer surface quality, at high f, resulted in a higher corrosion effect, and vice versa. Hence, f seemed to be a little more influent only in the highest range of the tested values.

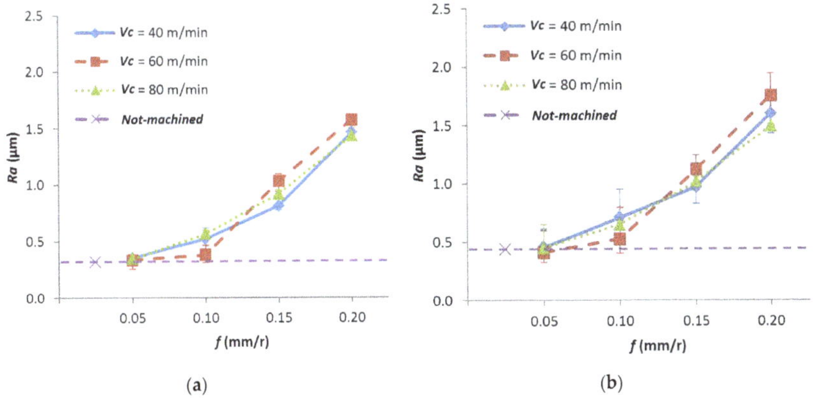

Figure 4. Average roughness (*Ra*) as a function of feed (f) and cutting speed (v_c), (**a**) before and (**b**) after a corrosion process.

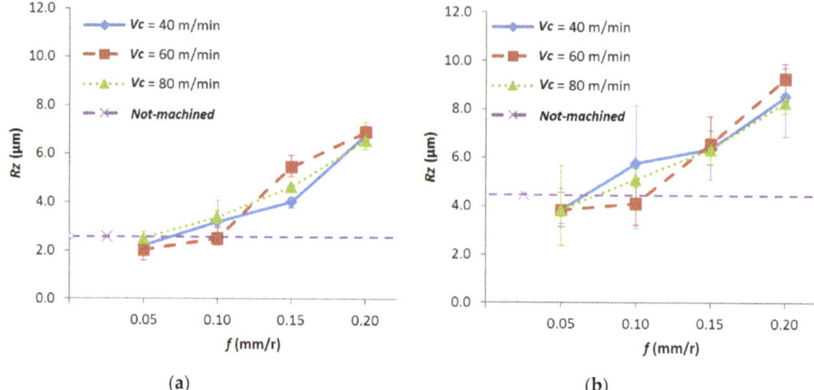

Figure 5. Maximum height of the profile (*Rz*) as a function of feed (*f*) and cutting speed (*v_c*), (**a**) before and (**b**) after a corrosion process.

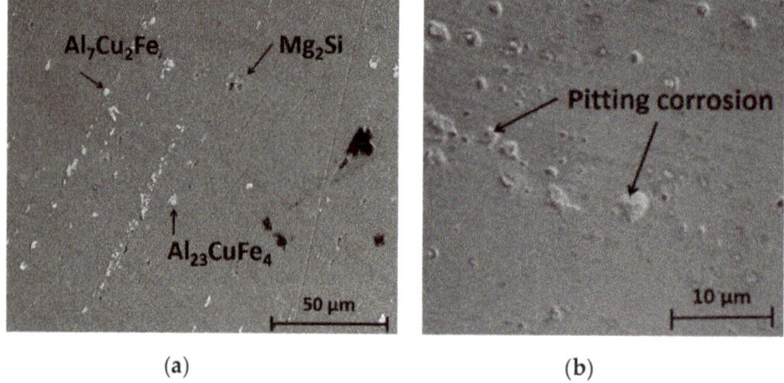

Figure 6. (**a**) SEM image of the microstructural composition of the UNS A97075 alloy tested (before machining and corrosion processes); (**b**) SEM image of the pitting corrosion effect on the machined surface.

Figure 7 shows the average microhardness values as a function of v_c and f, after and before the corrosion process, for the machined samples. In addition, the microhardness values for the non-machined samples are also plotted, as reference values.

In general, a surface microhardness (*HV*) increment was observed in all samples, after machining and before corrosion (Figure 7a), regardless of v_c and f. Its value was increased by 25% for a certain cutting parameter combination (f = 0.15 mm/r; v_c = 60 m/min), regarding the non-machined sample. This fact can be explained taking into account that machining resulted in very aggressive conditions (high pressures and temperatures) in the cutting area, both in the tool and the part. The compression forces gave rise to a plastic deformation of the machined surface [49], as well as microstructural alterations (finer grain structure), which resulted in strain hardening [29] (Figure 8).

Regarding the cutting speed influence on microhardness, Figure 7a shows very similar *HV* mean values for v_c = 40 and 80 m/min, whereas its mean value seemed to increase for v_c = 60 m/min, for the intermediate range of f (0.10 and 0.15 mm/r). In addition, higher dispersion in the experimental results was observed for this cutting parameter combination. With regard to f, the best results (the highest *HV*) were obtained for f = 0.10 and 0.15 mm/r. When these values were combined with the intermediate v_c tested value (60 m/min), *HV* reached a difference of 25% compared to other cutting speed values (40 and 80 m/min). For other f (0.05 and 0.20 mm/r) and v_c (40 and 80 m/min), a clear *HV* trend as a function of the cutting parameters was not observed.

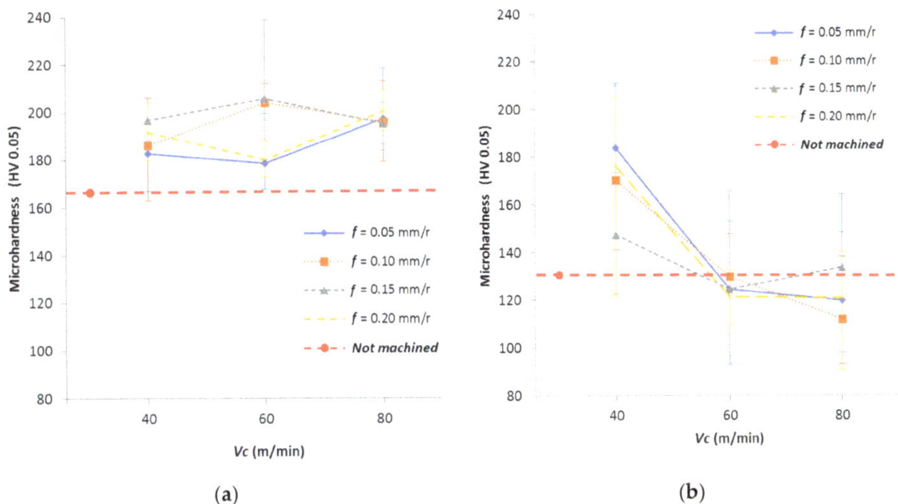

Figure 7. Surface microhardness (**a**) before corrosion and (**b**) after corrosion.

Therefore, the highest *HV* results were obtained when the intermediate range of cutting speed and feed values was applied. These results may be explained considering the effect that f and v_c produce on the surface microstructure. On one hand, f increments usually produce a higher plastic deformation on the machined surface due to mechanical effects (strain hardening), whereas the thermal effects are usually lower [49]. Therefore, a finer grain microstructure is obtained, giving rise to *HV* increments. On the other hand, v_c increments result in higher temperature in the cutting area [28]. The produced heat is removed from the cutting area by the chip, the tool, and the part. At high cutting speeds, most of the energy is dissipated trough the chip. For low cutting speeds (such as those used in this work), the dissipated energy by the chip becomes lower; thus, the dissipated heat through the tool and the part is increased. This fact results in a grain size increment and an *HV* reduction [29]. When v_c is increased (within this low range), this effect becomes more relevant.

Hence, for low f (0.05 mm/r) and v_c (40 m/min), mechanical effects predominated over thermal effects. The grain microstructure was finer than in the non-machined part (from 18.35 to 15.36 μm average grain size), and *HV* was higher (Figure 8c). For the intermediate range of f (0.10–0.15 mm/r) and v_c (60 m/min), the mechanical effects were stronger (Figure 8b). Thermal effects were increased but were not high enough to compensate for the mechanical effects. As a result, the grain microstructure was finer (14.04 μm average grain size) than for low f and v_c, and *HV* was increased.

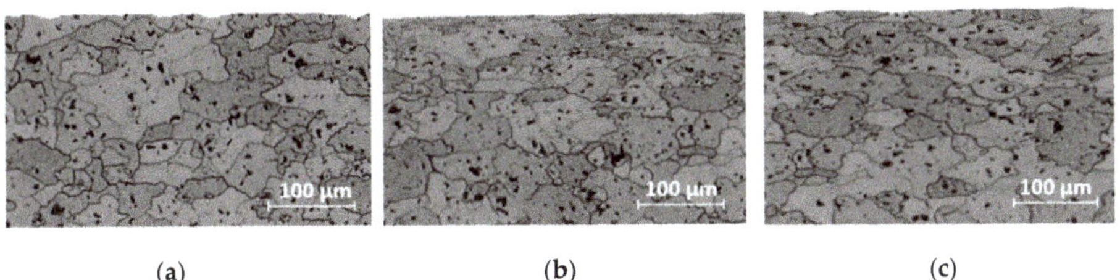

Figure 8. SEM images of the surface microstructure: (**a**) non-machined; (**b**) v_c = 60 m/min and f = 0.15 mm/r; (**c**) v_c = 40 m/min and f = 0.05 mm/r.

Finally, for the highest v_c (80 m/min), the thermal effects became more relevant and compensated for the mechanical effects. Therefore, the grain size was greater and *HV* was reduced, reaching a similar size to that for low f values. Consequently, mechanical and thermal effects were mixed, and the feed mechanical effects could not be neglected against the thermal effects of cutting speed, within the range of the tested v_c values. This fact made it more difficult to obtain a clear trend of HV as a function of the cutting parameters.

With regard to the HV results after corrosion, a general microhardness reduction was observed in the samples, both before and after machining (Figure 7b). In addition, a higher dispersion was obtained in all the range of the tested cutting parameters. The irregularities in the surface, due to pitting corrosion, resulted in a higher dependence on the surface point where HV was measured. Similar average values were obtained for $v_c = 60$ and 80 m/min and the sample before machining. The average value seemed to be higher only for $v_c = 40$ m/min. However, it was difficult to establish a general trend, given the high level of dispersion that the results showed. Therefore, the corrosion process negatively affected microhardness, but it was not possible to establish a direct relationship between cutting parameters, surface roughness, and microhardness after a corrosion process.

4. Conclusions

In this work, the cutting speed and feed influence on the microhardness of dry-turned UNS A97075-T6 alloy samples was analyzed. The results were compared with the non-machined samples, before and after a corrosion process, in order to study the possible connection between cutting parameters, microgeometrical deviations, microhardness, and corrosion.

The microgeometrical deviations, evaluated through Ra and Rz, showed a general trend to increase with feed. However, no significant modifications were observed as a function of cutting speed. The natural roughness, induced by the feed, was predominant over the cutting speed thermal effects (such as surface alterations originated by the indirect adhesion tool wear). These observations are in good agreement with previous research regarding parametric analysis on surface roughness for the 7000 aluminum alloys series.

In addition, no significate changes in Ra were observed after corrosion. However, an Rz increment was observed, regardless of feed and cutting speed. For the highest feed value (0.20 mm/r), this increment was slightly higher. Therefore, feed seemed to be more influent for the highest tested values. Additionally, the dispersion in the results seemed to be higher for Ra. This fact was more evident for Rz. The appearance of surface alterations, due to pitting corrosion, resulted in a higher dispersion of the maximum height of the surface profile (Rz). This effect was softer for Ra, because it was calculated as an average value of the surface profile deviations.

In general, a surface microhardness (HV) increment was observed in all samples, after machining and before corrosion, regardless of v_c and f. The highest microhardness was obtained for the intermediate cutting speed (60 m/min) and feed (0.10–0.15 mm/r) values. For low feed and cutting speed, mechanical effects predominated over thermal effects. This fact resulted in finer microstructure than in the non-machined part, giving rise to microhardness increments. For the intermediate range of feed and cutting speed, the mechanical effects were stronger. The thermal effects were increased, but not high enough to compensate for the mechanical effects. As a result, the grain microstructure was finer than for low feed and cutting speed. Finally, for the highest cutting speed (80 m/min), the thermal effects were more relevant and compensated for the mechanical effects, increasing the grain size and reducing the microhardness. Therefore, the mechanical effects, produced by feed, should not be neglected against the thermal effects, produced by cutting speed, within the low range of tested cutting speed.

A general microhardness reduction was observed in the samples after corrosion. However, a higher dispersion in the results was obtained in a wide range of cutting parameters analyzed, due to the irregularities in the surface produced by pitting corrosion. Therefore, the corrosion process negatively affected microhardness, but it was not possible to establish a direct relationship between cutting parameters, surface roughness, and microhardness after a corrosion process.

It is necessary to highlight that previous works in this regard usually focused on high-speed values of cutting speed, typically used on non-hybridized alloy. Under these conditions, the cutting speed is the most relevant cutting parameter and, therefore, thermal effects predominate over the mechanical ones. However, there is a gap in knowledge regarding the analysis of the possible influence of feed when low cutting speeds are used. Hence, this work covers this gap.

Finally, it is necessary to point out that this work is framed within a broader research line, currently in development, regarding the analysis of the cutting parameter influence on the surface integrity of UNS A97075 alloy, turned under dry conditions. This research line focuses on analyzing the relationship between cutting parameters, surface topography, and mechanical properties.

Author Contributions: S.M.-B. and F.J.T.V. conceptualized and designed the experiments; S.M.-B. performed the experiments; S.M.-B., F.J.T.V., C.B.G., and L.S.H. analyzed the data; S.M.-B., F.J.T.V., and C.B.G. wrote the paper; L.S.H. revised the paper. All authors have read and agreed to the published version of the manuscript.

Funding: This research received no external funding.

Acknowledgments: The authors thank the University of Malaga-Andalucia Tech Campus of International Excellence for its economic contribution to this paper.

Conflicts of Interest: The authors declare no conflicts of interest.

References

1. Morinaga, M. A Quantum Approach to Alloy Design. In *Aluminum Alloys and Magnesium Alloys*; Elsevier: Amsterdam, The Netherlands, 2019; pp. 95–130.
2. Campbell, F.C. *Manufacturing Technology for Aerospace Structural Materials*; Elsevier Science: Oxford, UK, 2006.
3. Benchergui, D.; Svoboda, C. *Aircraft Design*; Aerospace: Washington, DC, USA, 2012; Volume 50, p. 28.
4. Martín Béjar, S.; Trujillo Vilches, F.J.; Bermudo Gamboa, C.; Sevilla Hurtado, L. Parametric analysis of macro-geometrical deviations in dry turning of UNS A97075 (Al-Zn) alloy. *Metals* **2019**, *9*, 1141. [CrossRef]
5. Suraratchai, M.; Limido, J.; Mabru, C.; Chieragatti, R. Modelling the influence of machined surface roughness on the fatigue life of aluminium alloy. *Int. J. Fatigue* **2008**, *30*, 2119–2126. [CrossRef]
6. Hernández-González, L.W.; Pérez-Rodríguez, R.; Dumitrescu, L.; Montero-Sarmiento, R. Estudio de la influencia de los parámetros de corte en la integridad superficial y las desviaciones durante el fresado del acero AISI 1010. *Revista Tecnología en Marcha* **2015**, *28*, 26–41. [CrossRef]
7. M 'saoubi, R.; Outeiro, J.C.C.; Chandrasekaran, H.; Dillon, O.W.; Jawahir, I.S.S. A review of surface integrity in machining and its impact on functional performance and life of machined products. *Int. J. Sustain. Manuf.* **2008**, *1*, 203. [CrossRef]
8. Matsumoto, Y.; Hashimoto, F.; Lahoti, G. Surface integrity generated by precision hard turning. *CIRP Ann. -Manuf. Technol.* **1999**, *48*, 59–62. [CrossRef]
9. Abrão, A.M.; Ribeiro, J.L.S.; Davim, J.P. Surface Integrity. In *Machining of Hard Materials*; Springer Science & Business Media: New York, NY, USA, 2011; pp. 115–141.
10. Goncharenko, A.V. Aeronautical and Aerospace Material and Structural Damages to Failures: Theoretical Concepts. *Int. J. Aerosp. Eng.* **2018**, *2018*, 4126085. [CrossRef]
11. Benedicto, E.; Carou, D.; Rubio, E.M. Technical, Economic and Environmental Review of the Lubrication/Cooling Systems Used in Machining Processes. *Procedia Eng.* **2017**, *184*, 99–116. [CrossRef]
12. Kopac, J.; Pusavec, F.; Krolczyk, G. Cryogenic machining, surface integrity and machining performance. *Arch. Mater. Sci. Eng.* **2015**, *71*, 83–93.
13. Krolczyk, G.M.; Maruda, R.W.; Krolczyk, J.B.; Wojciechowski, S.; Mia, M.; Nieslony, P.; Budzik, G. Ecological trends in machining as a key factor in sustainable production—A review. *J. Clean. Prod.* **2019**, *218*, 601–615.13. [CrossRef]
14. Saleem, W.; Zain-ul-abdein, M.; Ijaz, H.; Salmeen Bin Mahfouz, A.; Ahmed, A.; Asad, M.; Mabrouki, T. Computational Analysis and Artificial Neural Network Optimization of Dry Turning Parameters—AA2024-T351. *Appl. Sci.* **2017**, *7*, 642. [CrossRef]
15. Do, T.-V.; Hsu, Q.-C. Optimization of Minimum Quantity Lubricant Conditions and Cutting Parameters in Hard Milling of AISI H13 Steel. *Appl. Sci.* **2016**, *6*, 83. [CrossRef]

16. Vilches, F.; Hurtado, L.; Fernández, F.; Gamboa, C. Analysis of the Chip Geometry in Dry Machining of Aeronautical Aluminum Alloys. *Appl. Sci.* **2017**, *7*, 132. [CrossRef]

17. Kavitha, K.; Vijayan, R.; Sathishkumar, T. Fibre-metal laminates: A review of reinforcement and formability characteristics. *Mater. Today Proc.* **2019**. [CrossRef]

18. Lu, J.; Song, Y.; Hua, L.; Zheng, K.; Dai, D. Thermal deformation behavior and processing maps of 7075 aluminum alloy sheet based on isothermal uniaxial tensile tests. *J. Alloys Compd.* **2018**, *767*, 856–869. [CrossRef]

19. Syed, A.K.; Zhang, X.; Moffatt, J.E.; Fitzpatrick, M.E. Effect of temperature and thermal cycling on fatigue crack growth in aluminium reinforced with GLARE bonded crack retarders. *Int. J. Fatigue* **2017**, *98*, 53–61. [CrossRef]

20. Caruso, S.; Rotella, G.; Del Prete, A.; Umbrello, D. Finite Element Modeling of Microstructural Changes in Hard Machining of SAE 8620. *Appl. Sci.* **2019**, *10*, 121. [CrossRef]

21. Meng, X.; Lin, Z.; Wang, F. Investigation on corrosion fatigue crack growth rate in 7075 aluminum alloy. *Mater. Des.* **2013**, *51*, 683–687. [CrossRef]

22. Yang, X.W.; Li, W.Y.; Li, H.Y.; Yao, S.T.; Sun, Y.X.; Mei, L. Microstructures and microhardness for sheets and TIG welded joints of TA15 alloy using friction stir spot processing. *Trans. Nonferrous Met. Soc. China* **2018**, *28*, 55–65. [CrossRef]

23. Privezentsev, D.; Zhiznyakov, A.; Kulkov, Y. Analysis of the Microhardness of Metals Using Digital Metallographic Images. *Mater. Today Proc.* **2019**, *11*, 325–329. [CrossRef]

24. Jirapattarasilp, K.; Kuptanawin, C. Effect of Turning Parameters on Roundness and Hardness of Stainless Steel: SUS 303. *Procedia-Soc. Behav. Sci.* **2012**, *3*, 160–165. [CrossRef]

25. Sasahara, H. The effect on fatigue life of residual stress and surface hardness resulting from different cutting conditions of 0.45%C steel. *Int. J. Mach. Tool Manuf.* **2004**, *45*, 131–136. [CrossRef]

26. Akkurt, A. The effect of cutting process on surface microstructure and hardness of pure and Al 6061 aluminium alloy. *Eng. Sci. Technol. Int. J.* **2015**, *18*, 303–308. [CrossRef]

27. Kurkute, V.; Chavan, S.T. Modeling and Optimization of surface roughness and microhardness for roller burnishing process using response surface methodology for Aluminum 63400 alloy. *Procedia Manuf.* **2018**, *20*, 542–547. [CrossRef]

28. Surya Sundara Rao, K.; Viswanath Allamraju, K. Effect on Micro-Hardness and Residual Stress in CNC Turning of Aluminium 7075 Alloy. *Mater. Today Proc.* **2017**, *4*, 975–981. [CrossRef]

29. Rotella, G.; Dillon, O.W.; Umbrello, D.; Settineri, L.; Jawahir, I.S. Finite element modeling of microstructural changes in turning of AA7075-T651 Alloy. *J. Manuf. Process.* **2013**, *15*, 87–95. [CrossRef]

30. Campbell, C.E.; Bendersky, L.A.; Boettinger, W.J.; Ivester, R. Microstructural characterization of Al-7075-T651 chips and work pieces produced by high-speed machining. *Mater. Sci. Eng. A* **2006**, *430*, 15–26. [CrossRef]

31. Sravanthi, M.; Manjunatha, K.G. Corrosion Studies of As Casted and Heat Treated Aluminium-7075 Composites. *Mater. Today Proc.* **2018**, *5*, 22581–22594. [CrossRef]

32. Dong, H. *Surface Engineering of Light Alloys: Aluminium, Magnesium and Titanium Alloys*; CRC Press: Amsterdam, The Netherlands, 2010.

33. Guérin, M.; Alexis, J.; Andrieu, E.; Blanc, C.; Odemer, G. Corrosion-fatigue lifetime of Aluminium-Copper-Lithium alloy 2050 in chloride solution. *Mater. Des.* **2015**, *87*, 681–692. [CrossRef]

34. Liu, B.; Zhang, X.; Zhou, X.; Hashimoto, T.; Wang, J. The corrosion behaviour of machined AA7150-T651 aluminium alloy. *Corros. Sci.* **2017**, *126*, 265–271. [CrossRef]

35. Dong, P.; Sun, D.; Wang, B.; Zhang, Y.; Li, H. Microstructure, microhardness and corrosion susceptibility of friction stir welded AlMgSiCu alloy. *Mater. Des.* **2014**, *54*, 760–765. [CrossRef]

36. Wang, X.; Nie, M.; Wang, C.T.; Wang, S.C.; Gao, N. Microhardness and corrosion properties of hypoeutectic Al-7Si alloy processed by high-pressure torsion. *Mater. Des.* **2015**, *83*, 193–202. [CrossRef]

37. Donatus, U.; Viveiros, B.V.G.; de Alencar, M.C.; de Ferreira, R.O.; Milagre, M.X.; Costa, I. Correlation between corrosion resistance, anodic hydrogen evolution and microhardness in friction stir weldment of AA2198 alloy. *Mater. Charact.* **2018**, *144*, 99–112. [CrossRef]

38. Sherafat, Z.; Paydar, M.H.; Ebrahimi, R. Fabrication of Al7075/Al, two phase material, by recycling Al7075 alloy chips using powder metallurgy route. *J. Alloys Compd.* **2009**, *487*, 395–399. [CrossRef]

39. Zhenchao, Y.; Yang, X.; Yan, L.; Jin, X.; Quandai, W. The effect of milling parameters on surface integrity in high-speed milling of ultrahigh strength steel. *Ann. DAAAM Proc.* **2019**, *30*, 753–757. [CrossRef]

40. Yue, T.M.; Yu, J.K.; Man, H.C. Corrosion behavior of aluminum 2009/SiC composite. *Mater. Today Proc.* **2009**, *2*, 1069–1072.

41. Bienvenido, R.; Díaz Vázquez, J.E.; Botana, J.; Cano, M.J.; Marcos, M. Preliminary study of the influence of machining conditions in the response to corrosion of UNS-A92024 alloy. *Proc. Adv. Mater Res.* **2010**, *107*, 117–121. [CrossRef]

42. International Standaring Organization. *ISO 1302:2002. Geometrical Product Specifications (GPS)—Indication of Surface Texture in Technical Product Documentation*; International Standaring Organization: Geneva, Switzerland, 2002.

43. International Standaring Organization. *ISO 6507-1:2018. Metallic Materials—Vickers Hardness Test-Part 1: Test Method*; International Standaring Organization: Geneva, Switzerland, 2018.

44. Tank, K.; Shetty, N.; Panchal, G.; Tukrel, A. Optimization of turning parameters for the finest surface roughness characteristics using desirability function analysis coupled with fuzzy methodology and ANOVA. *Mater. Today Proc.* **2018**, *5*, 13015–13024. [CrossRef]

45. Trujillo, F.J.; Marcos Bárcena, M.; Sevilla, L. Experimental Prediction Model for Roughness in the Turning of UNS A97075 Alloys. *Mater. Sci. Forum* **2014**, *797*, 59–64. [CrossRef]

46. Torres, A.; Puertas, I.; Luis, C.J. Surface roughness analysis on the dry turning of an Al-Cu alloy. *Procedia Eng.* **2015**, *132*, 537–544. [CrossRef]

47. Rubio, E.M.; Camacho, A.M.; Sánchez-Sola, J.M.; Marcos, M. Surface roughness of AA7050 alloy turned bars: Analysis of the influence of the length of machining. *Proc. J. Mater. Proces. Technol.* **2005**, *162–163*, 682–689. [CrossRef]

48. Bandeira, R.M.; van Drunen, J.; Garcia, A.C.; Tremiliosi-Filho, G. Influence of the thickness and roughness of polyaniline coatings on corrosion protection of AA7075 aluminum alloy. *Electrochim. Acta* **2017**, *240*, 215–224. [CrossRef]

49. Ajithkumar, J.; Xavior, M.A. Cutting Force and Surface Roughness Analysis During Turning of Al 7075 Based Hybrid Composites. *Procedia Manuf.* **2019**, *30*, 180–187. [CrossRef]

MDPI

St. Alban-Anlage 66

4052 Basel

Switzerland

Tel. +41 61 683 77 34

Fax +41 61 302 89 18

www.mdpi.com

Applied Sciences Editorial Office

E-mail: applsci@mdpi.com

www.mdpi.com/journal/applsci

9 783039 436217